Springer Complexity

Springer Complexity is a publication program, cutting across all traditional disciplines of sciences as well as engineering, economics, medicine, psychology and computer sciences, which is aimed at researchers, students and practitioners working in the field of complex systems. Complex Systems are systems that comprise many interacting parts with the ability to generate a new quality of macroscopic collective behavior through self-organization, e.g., the spontaneous formation of temporal, spatial or functional structures. This recognition, that the collective behavior of the whole system cannot be simply inferred from the understanding of the behavior of the individual components, has led to various new concepts and sophisticated tools of complexity. The main concepts and tools – with sometimes overlapping contents and methodologies – are the theories of self-organization, complex systems, synergetics, dynamical systems, turbulence, catastrophes, instabilities, nonlinearity, stochastic processes, chaos, neural networks, cellular automata, adaptive systems, and genetic algorithms.

The topics treated within Springer Complexity are as diverse as lasers or fluids in physics, machine cutting phenomena of workpieces or electric circuits with feedback in engineering, growth of crystals or pattern formation in chemistry, morphogenesis in biology, brain function in neurology, behavior of stock exchange rates in economics, or the formation of public opinion in sociology. All these seemingly quite different kinds of structure formation have a number of important features and underlying structures in common. These deep structural similarities can be exploited to transfer analytical methods and understanding from one field to another. The Springer Complexity program therefore seeks to foster cross-fertilization between the disciplines and a dialogue between theoreticians and experimentalists for a deeper understanding of the general structure and behavior of complex systems.

The program consists of individual books, books series such as "Springer Series in Synergetics", "Institute of Nonlinear Science", "Physics of Neural Networks", and "Understanding Complex Systems", as well as various journals.

Understanding Complex Systems

Series Editor

J.A. Scott Kelso
Florida Atlantic University
Center for Complex Systems
Glades Road 777
Boca Raton, FL 33431-0991, USA

Understanding Complex Systems

Future scientific and technological developments in many fields will necessarily depend upon coming to grips with complex systems. Such systems are complex in both their composition (typically many different kinds of components interacting with each other and their environments on multiple levels) and in the rich diversity of behavior of which they are capable. The Springer Series in Understanding Complex Systems series (UCS) promotes new strategies and paradigms for understanding and realizing applications of complex systems research in a wide variety of fields and endeavors. UCS is explicitly transdisciplinary. It has three main goals: First, to elaborate the concepts, methods and tools of self-organizing dynamical systems at all levels of description and in all scientific fields, especially newly emerging areas within the Life, Social, Behavioral, Economic, Neuro- and Cognitive Sciences (and derivatives thereof); second, to encourage novel applications of these ideas in various fields of Engineering and Computation such as robotics, nanotechnology and informatics; third, to provide a single forum within which commonalities and differences in the workings of complex systems may be discerned, hence leading to deeper insight and understanding. UCS will publish monographs and selected edited contributions from specialized conferences and workshops aimed at communicating new findings to a large multidisciplinary audience.

R.R. McDaniel, Jr. D.J. Driebe (Eds.)

Uncertainty and Surprise in Complex Systems

Questions on Working with the Unexpected

With 28 figures and 8 tables

Springer

Reuben R. McDaniel, Jr.
The University of Texas at Austin
McCombs School of Business
MSIS Department B6500
1 University Station
Austin, TX 78712-0212
USA

Dean J. Driebe
The University of Texas at Austin
I. Prigogine Ctr. for Stat. Mech. & Comp. Sys.
Department of Physics C1609
1 University Station
Austin, TX, 78712-0264
USA

ISSN 1860-0832
ISBN 978-3-642-06272-8 e-ISBN 978-3-540-32372-3

Springer is a part of Springer Science+Business Media
springeronline.com

Printed in The Netherlands

Cover-Design: Erich Kirchner, Heidelberg

Printed on acid-free paper 89/3141/Yu – 5 4 3 2 1 0

Preface and Acknowledgements

The papers in this volume were first presented at the conference, "Uncertainty and Surprise: Questions on Working with the Unexpected and Unknowable," held April 10-12, 2003 at the McCombs School of Business at The University of Texas at Austin. Sponsors of the conference were The Plexus Institute and The Ilya Prigogine Center for Studies in Statistical Mechanics and Complex Systems. The Hana Foundation provided generous financial support that made the conference possible. Generous financial support was received from the IC2 Institute, the Herb Kelleher Center for Entrepreneurship and, the McCombs School of Business all located at The University of Texas at Austin. Additional financial support was provided by the IBM Center for Advanced Studies.

This conference would never have been held without the leadership and support of Curt Lindberg, President and Henri Lipmanowicz, Chair of the Board of the Plexus Institute. They were both involved right from the initial conceptual efforts to the opening of the conference through the preparation of these proceedings. Bob Shapiro, a member of the Board of the Plexus Institute, was an early contributor to the planning for the conference. The Plexus Institute is a leader in the promotion of a complexity science framework for examining a multitude of issues. The leadership of the Institute saw this conference as a way to bring ideas from complexity science to bear on our understanding of uncertainty and surprise. Curt Lindberg, in particular, kept his interest high, was key in contacting and making arrangements with speakers, and was a source of encouragement for all associated with the conference.

Ilya Prigogine, Director and Linda Reichl, Associate Director of the Prigogine Center both provided significant leadership as the ideas for the conference developed. The Prigogine Center is a leader in the study of Chaos Theory and Complexity Science and they saw this conference as a way to extend the work of the Center in a meaningful way. Linda Reichl participated in several meetings of the planning committee and lent considerable expertise to the planning of the conference. The Center was also quite instrumental in attracting both speakers and participants to the conference.

John Butler, Director of IC2 Institute, was extremely influential in raising the financial support for the conference. He was able to see the value of such a conference and to convince others of its value.

Several people from the staffs of the McCombs School, the Plexus Institute and the Prigogine Center provided crucial support in making the logistics of the conference work. Nan Watkins, from the McCombs School was determined that the conference would be a spectacular success and she worked tirelessly to make it so. The organizational skills of Annie Harding from the Prigogine Center contributed to the success of the meeting. The staff of the Plexus Institute handled the administrative arrangements for speakers and the registration for the conference without a hitch. Special thanks to Scott Kelso, editor of the Springer Series in Understanding Complex Systems for accepting these conference proceedings as a volume in this Series and for his encouragement during its preparation.

Conference speakers were most helpful in working with the conference staff and in preparing material for publication in this volume. They represent a broad spectrum of academic disciplines and organizational roles indicating the breadth and depth of interest in the conference theme. They came to the conference, presented well-thought-out papers, and shared in our discussions throughout the conference. Clearly their intellectual efforts were the source of any accomplishments we enjoyed.

Last but not least, we wish to thank the more than one hundred conference participants who came to the conference and gave of their time and energy for its success. Without their participation, the conference would not have worked and these proceedings would never have come into existence. We hope that they enjoyed the conference and that they believe that they have a deeper understanding of uncertainty and surprise as the result of their attendance.

Reuben R. McDaniel, Jr.
Dean J. Driebe

Contents

Contributing Authors

Peter M. Allen, Professor of Evolutionary Complex Systems, Head of Complex Systems Management Centre, School of Management, Cranfield University, UK, p.m.allen@cranfield.ac.uk

Erich K. Baier, Director, UNIX Hardware Development, IBM Corporation, Austin, TX, ebaier@us.ibm.com

James S. Baldwin, Advanced Manufacturing Research Centre, Department of Mechanical Engineering, University of Sheffield, UK

James W. Begun, James A. Hamilton Term Professor, Department of Healthcare Management, Carlson School of Management, University of Minnesota, jbegun@csom.umn.edu

Benjamin F. Crabtree, Professor and Research Director, Department of Family Medicine, Robert Wood Johnson Medical School, University of Medicine and Dentistry of New Jersey, One Robert Wood Johnson Place, MEB 242, New Brunswick, New Jersey 08903-0029, crabtrbf@umdnj.edu

Dean J. Driebe, Research Scientist, The Ilya Prigogine Center for Studies in Statistical Mechanics and Complex Systems, The University of Texas at Austin, dean@physics.utexas.edu

James S. Dyer, Professor, The Fondren Foundation Centennial Chair in Business, MSIS Department, McCombs School of Business, The University of Texas at Austin, 1 University Station B6500, Austin, TX 78712-0212, j.dyer@mail.utexas.edu

June Holley, President/CEO, Appalachian Center for Economic Networks (ACEnet), juneh@acenetworks.org

Michelle E. Jordan, Editor, MSIS Department, McCombs School of Business, The University of Texas at Austin, majordan@austin.isd.tenet.edu

Amer A. Kaissi, Assistant Professor, Department of Healthcare Administration, Trinity University, One Trinity Place, #58, San Antonio, TX 78212, amer.kaissi@trinity.edu

J. A. Scott Kelso, Complex Systems and Brain Sciences, Department of Psychology, Florida Atlantic University, 777 Glades Rd., Boca Raton, Florida, 33431-0991, kelso@ccs.fau.edu

Larry S. Liebovitch, Professor, Florida Atlantic University, Center for Complex Systems and Brain Sciences, Center for Molecular Biology and Biotechnology, Department of Psychology, Florida Atlantic University, 777 Glades Rd., Boca Raton, Florida, 33431-0991, liebovitch@clifford.ccs.fau.edu

Reuben R. McDaniel, Jr., Professor, Charles and Elizabeth Prothro Regents Chair, MSIS Deptartment, McCombs School of Business, The University of Texas at Austin, 1 University Station B6500, Austin, TX 78712-0212, reuben.mcdaniel@mccombs.utexas.edu

John C. Peirce, Former Chief Academic Officer, Banner Good Samaritan Medical Center, Phoenix, AZ, jcpeirce@comcast.net

Tom Petzinger, Chairman and CEO, LaunchCyte LLC., A Bio-Informatics Network, 100 Technology Drive, Suite 440, Pittsburgh, PA 15219, tom@launchcyte.com

Ilya Prigogine, Regental Professor and Ashbel Smith Professor of Physics and Chemical Engineering and Director, Centre for Studies in Statistical Mechanics and Complex Systems, The University of Texas at Austin, and Director, International Solvay Institutes for Physics and Chemistry, Brussels, Belgium (deceased)

Linda E. Reichl, Professor of Physics and Acting Director, Center for Studies in Statistical Mechanics and Complex Systems, Physics Department, The University of Texas at Austin, 1 University Station C1600, Austin, TX 78712, reichl@physics.utexas.edu

Mark Strathern, Complex Systems Management Centre, Cranfield University, UK

James H. Taylor, President and CEO, University of Louisville Hospital, Louisville, KY 40202, jtaylor@ulh.org

Karl E. Weick, Rensis Likert College Professor of Organizational Behavior and Psychology, Professor of Psychology, Department of Psychology, University of Michigan, 525 East University, Ann Arbor, MI 48109-1109, karlw@bus.umich.edu

Bruce J. West, Mathematics Division, U.S. Army Research Office, Research Triangle Park, NC 27709, westb@aro.arl.army.mil

Section I Introduction to Uncertainty and Surprise

1 Uncertainty and Surprise: An Introduction

Reuben R. McDaniel, Jr.[1] and Dean J. Driebe[2]

Much of the traditional scientific and applied scientific work in the social and natural sciences has been built on the supposition that the unknowability of situations is the result of a lack of information. This has led to an emphasis on uncertainty reduction through ever-increasing information seeking and processing, including better measurement and observational instrumentation. Pending uncertainty reduction through better information, efforts are devoted to uncertainty management and hierarchies of controls. A central goal has been the avoidance of surprise.

Complex systems research and the study of chaotic dynamics have demonstrated that unpredictability and surprise are fundamental aspects of the world around us. These fields have also injected a narrative, multi-faceted view into our description of the physical world that was lacking in the classical atemporal description.

Instead of the conception of natural and social systems as machines whose Newtonian-like dynamics need to be uncovered and then controlled, we now have the viewpoint that they are self-organizing systems whose properties emerge from the nonlinear interactions of agents.

In particular, when we have studied the nature of complex systems we see that uncertainty is fundamental and not simply the result of lack of information. Therefore, we are faced with real limitations on our ability to understand these systems through better information or measurement technologies. But it is unclear what the best strategies might be for developing an increased understanding of complex systems, the fundamental uncertainty that they present and the surprises that occur as they unfold. Because of this situation, a conference was held, sponsored by the Prigogine Center at The University of Texas at Austin and the Plexus Institute to explore strategies for understanding uncertainty and surprise. The purposes of this conference were to address:

- The impacts of these new perspectives on the research agenda of the natural and social sciences,
- The relevance of these perspectives for practitioners/executives in public policy and management,
- The new research and practice questions at the individual, organizational and community levels that need to be explored for increasing our understanding of fundamental uncertainties and for increasing our ability to cope with these uncertainties.

---

[1] Charles and Elizabeth Prothro Regents Chair, McCombs School of Business, The University of Texas at Austin, reuben.mcdaniel@mccombs.utexas.edu.

[2] Research Scientist, The Ilya Prigogine Center for Studies in Statistical Mechanics and Complex Systems, The University of Texas at Austin, dean@physics.utexas.edu.

These purposes required an unconventional conference approach. Because of the nonlinear feedback loops between these areas of concern, it was imperative that critical thinkers from a variety of perspectives interact face to face. We brought together leading scholars, with expertise in complex systems, from the natural, physical, and social sciences, and leaders from a wide variety of organizations (i.e., government, business, education, health care, military and community). The basic goal was to identify and clarify questions rather than to proclaim answers.

1.1 Complex Adaptive Systems (CAS)

It will be helpful to articulate the understandings of complexity science and complex adaptive systems that guided our considerations as we began to explore the conference topics and to identify people who might be interested in joining us. Complexity science offers new ways to think about organizations that enables one to have new insights about their nature and about their functioning (Begun 1985; Capra 1996). Complexity science transcends traditional disciplines and has been a source of new insights in physics, biology, geology, cosmology as well as the social sciences (Fontana and Ballati 1999). As noted by Mainzer (1996:272), "The crucial point of the complex systems approach is that from a macroscopic point of view the development of political, social or cultural order is not only the sum of single intentions, but the collective result of nonlinear interactions." When we look at the world through the lens of more conventional science it may seem as though order is unnatural because the orderly arrangements of elements seems so unlikely. One of the aims of complexity science is to explain why ordering can appear spontaneously in the universe (Johnson 1995; Kauffman 1993).

Complexity science deals with systems that are characterized by nonlinear dynamics and emergent properties. One of the types of systems often studied in complexity science are known as Complex Adaptive Systems (CAS). CAS are characterized by diverse agents interacting with each other and the system is capable of undergoing spontaneous self-organization (Cilliers 1998). The state of a complex adaptive system as a whole is irreducible to a linear superposition of the states of its constituent elements. The essence of the study of complex adaptive systems is in the study of patterns and relationships, and in the search for characteristics of systems far from equilibrium rather that at the point of balanced stability (Capra 1996). Complexity science looks not only at the parts, but at the whole in an effort to gain a deeper, qualitatively different understanding of phenomena. Complexity results from the interactions between the components of a system (Cilliers 1998) and characteristics at one level cannot be understood from knowledge of characteristics at other levels (Holland 1998; Newman 1996).

Complex adaptive systems are made up of a large number of agents that are information processors (Waldrop 1992; Holland 1998). These agents may be nerve cells, computer programs, individuals, or firms and each may be considered as a CAS itself. The one characteristic that these agents all share is that they can process information and react to changes in that information (Casti 1997). Agents have the capacity to exchange information among themselves and with their environment and to adjust their own behavior as a function of information they process (Holland 1995).

CAS agents are diverse from each other (Kauffman 1995; Coleman 1999). This diversity is critical to the ability of the CAS to function because diversity is a source of novelty and adaptability. Although agents are elements in their own right, and are often CAS themselves, it is also true that agents at any one level in a CAS serve as building blocks for agents at a higher level (Waldrop 1992). Different agents take different roles as the dynamic of the system unfolds. "CAS are constantly revising and rearranging their building blocks as they gain experience" (Waldrop 1992:146).

While a diverse set of agents is necessary for a CAS, it is not sufficient. In fact, the essence of a CAS is captured in the relationships among agents, rather than in the agents themselves. As scientists approach questions about organizations, their questions tend to refer to systems where there are a great many interdependent agents interacting with each other in a great many ways (Waldrop 1992). Relationships among agents in a CAS are non-linear in nature and the effect of any one agent's activity can feed back on itself as well as influence other agents.

1.2 Self-Organization

"Self-organization is the spontaneous emergence of new structures and new forms of behavior in open systems far from equilibrium, characterized by internal feedback loops and described mathematically by nonlinear equations" (Capra 1996:85). The structure and form is not simply externally imposed from some hierarchical controller. Rather, structure and form are a function of patterns of relationships among agents and interactions of these agents with their environment (Cilliers 1998; Mainzer 1996).

Two examples that have been often used for illustrating the self-organizing properties of CASs are the flocking of birds and the schooling of fish. In neither case is there some "smart" bird or fish that "gets things organized" (Callen and Shapero 1974). Rather the pattern of organization develops from local interactions among agents, with agents following very simple rules. This phenomenon of self-organization has been used to better understand how colonies of ants seek food and organize their living arrangements (Bossomaier and Green 1998).

When one observes order in a system, one is tempted to assert that the order must come from some intentionality on the part of some external controller. Complexity science teaches us that order in a system may well be a result of the properties of the system itself (Nicolis and Prigogine 1989; Kauffman 1993). Order is a result of nonlinear interactions and the capacity for self-organization is a function of (among other things) the number of connections among agents and the intensity of these connections. Too many connections may lead to behavior that never settles into any recognizable pattern of self-organization. While on the other hand, too few connections may lead to frozen behavior rather than dynamical self-organization. Kauffman (1995:84) expresses the importance of this observation as follows, "Our intuitions about the requirements for order have, I contend, been wrong for millennia. We do not need careful construction; we do not require crafting. We require only that extremely complex webs of interacting elements are sparsely coupled."

1.3 Emergence

"Emergence is above all a product of coupled, context-dependent interactions. Technically these interactions and the resulting system are *nonlinear*: The behavior of the overall system cannot be obtained by *summing* the behaviors of the constituent parts" (Holland 1998:122, italics in original). The global characteristics of the CAS arise from characteristics of agents and their relationships but are not reducible to these characteristics (Casti 1997). Emergence is the word we use to describe novelty and surprise in CASs (Goldstein 1999). The properties of an emergent system cannot be ascertained by observing the properties of lower level agents or subsystems. The unpredictability of emergent systems is fundamental. What is the outcome of emergence in CAS? There are new patterns of relationships among agents and these modify the self-organizing characteristics of CAS. These new patterns emerge from the nonlinear relationships among agents and the rules that constrain agents. Emergence is not the same as serendipitous novelty such as patterns of raindrops on a window pane but is the result of nonlinear dynamics generating new properties at the macro level of analysis (Goldstein 1999; Holland 1998).

1.4 Co-evolution

CAS consist of agents interacting in a nonlinear fashion such that the system self-organizes and emerges in a dynamic fashion. But the CAS does not simply change; it often changes the world around it. There is co-evolution of the CAS and its environment such that each fundamentally influences the development of the other (Kauffman 1993, 1995; McKelvey 1999). Agents do not simply adapt to the environment and each other. They co-evolve with each other and with the environment in a constant dance of change.

Kauffman (1993, 1995) has suggested that CASs exist in "fitness landscapes" and that each seeks a point of maximum fitness with its environment. Managers have often considered the need for their organizations to adapt to the environment but when they consider that every adaptive move creates another move by another organization or set of organizations, they then can see that adaptation is not sufficient. Each CAS acts based on local information, seeking to continuously improve its fit with its environment and, therefore, usually can only achieve some local optimum. In the process of achieving this position, each CAS changes the landscape for itself and for all other CAS in the system. As explained by Kauffman

> In a co-evolutionary system, we need to represent the fact that both the fitness and the fitness landscape of each species are a function of the other species. Thus, in general, it is necessary to couple the rugged fitness landscape for each species, such that an adaptive move by one species projects onto the fitness landscapes of the other species and alters those fitness landscapes more or less profoundly. (1993:243)

For CAS, the property of co-evolution signals limits in their developmental processes. Agents have conflicting constraints within themselves and among neighboring agents and because so many of the constraints are in conflict, compromise and cooperation lead to workable solutions rather than to some grand, superb solution (Kauffman 1995). The dynamics of the situation mean that you can't "get it right" in some global sense. Rather, because of the emerging properties of each agent and of each CAS, the "goodness" of CAS adaptation to its environment is a moving target.

1.5 Surprise

Traditionally, people in organizations have viewed surprises as unwelcome and generally dysfunctional occurrences, prompting actions to avoid or manage them. Bounded rationality (Simon 1991), lack of information, tight coupling and interdependence of system components (Perrow 1984) have been seen as major contributors to surprise and it is assumed that surprises can be and should be avoided with more knowledge, better planning, and/or better systems design (Yourstone and Smith 2002). The goal in the face of potential surprise is to create more reliable and predictable systems through quality control, planning, and standardization and/or to manage the unexpected in ways that reduce potential damage (Weick and Sutcliff 2001).

Certainly, some organizational surprise is due to lack of information or information processing capacity. When, however, organizations are recognized as Complex Adaptive Systems (CAS), surprise is not necessarily the result of bounded rationality, limited information or system design, but often is the result of the fundamental nature of the system in question. Complexity theory suggests that much surprise is inevitable because it is part of the natural order of things and cannot be avoided, eliminated, or controlled.

It might be helpful to give some very specific examples of surprises taken from the health care industry (McDaniel, Jordan and Fleeman 2003). While these observations may no longer be "surprises," they certainly were when they first were made and responses to each were, at one time, quite controversial.

A "steady heart beat" is seen as a dysfunction rather than a desired state and variability in cardiac rhythms is seen as a sign of good health (Goldberger 1997).

Participation of clinicians in hospital strategic decision-making is found to be more helpful in terms of bottom line performance than the participation of middle managers (Ashmos, Huonker and McDaniel 1998).

Even after purchasing costly systems designed to assist them in delivering preventive care, physicians ignored the protocols, the systems found their way to shelves in the back room, and the physician delivery of preventive care is found to be much more complex than previously suspected (Crabtree et al. 1998).

Organizations engaged in fast decision-making are discovered to use just as much information as slow decision makers but information of a different kind (Eisenhardt 1990).

Nursing homes that want to improve quality can use RN participation to make improvements without significantly increasing costs (Anderson and McDaniel 1999).

Surprises such as those noted above, are viewed in a multitude of ways as a function of the framework used (e.g., Izard 1977; Plutchik 1984; Mandler 1975; Lazarus 1991; Roseman 1984). Often, surprise is seen as an unfavorable deviation from past experiences and the organization makes an effort to go back in time. People often normalize surprise and deny its existence. Observers often attempt to enact surprise away so that they can know what to do and so that they are not confused by new information. Surprise in organizational life is predominantly viewed from a negative perspective implying a failure or mistake and a threat to organizational reliability and this has significant consequences organizational action. However, surprise could be seen in a positive light, as an opportunity rather than a threat, making organizational responses to surprise significantly different (Dutton and Jackson 1987).

Insights from complexity science show that the natural state of things is not a state of equilibrium:

> New opportunities are always being created by the system. And that, in turn, means that it's essentially meaningless to talk about a complex adaptive system being in equilibrium: the system can never get there. In fact, if the system ever does reach equilibrium, it isn't just stable. It's dead. (Waldrop 1992:147)

When the managers of organizations recognize the inevitable unpredictability of the unfolding of CAS over time, "surprise" takes on a different meaning and thus results in different kinds of responses and possibilities. For example, if the assumption is that the surprise happened because of the lack of sufficient information or speed, our response is likely to be a cry for bigger and better computers and more data collection; if bounded rationality is assumed to be causing the surprise, we will ask for more care, more vigilance, while pointing our fingers in search of who is to blame. However, if the assumption is that surprise often arises as the result of the fundamental unknowability of the world, we open the door for creative, innovative approaches without the mark of blame and failure. We change our relationship with the unexpected and resilience becomes a quality that is essential for effective management. Surprise seen negatively (as error) often leads to a search for reliability where as surprise seen positively (as opportunity) can lead to a search for new approaches to situations (McDaniel, Jordan and Fleeman

2003). "You have a system exploring its way into an immense space of possibilities, with no realistic hope of ever finding the single 'best' place to be. All evolution can do is look for improvements, not perfection" (Waldrop 1992:167).

1.6 Summary and Conclusions

People have tried historically to avoid the unexpected so that they can gain control over systems. Complexity science teaches us that there are elements to the unexpected that are unavoidable because they are in the very nature of the system dynamic. Of course there is uncertainty associated with ignorance. But even this uncertainty can benefit from a new perspective. We recognize that some surprises may be threats, but all surprises are threats if you enact them as threats. A different attitude toward the unexpected leads to a different approach, one that enables healthier responses. Uncertainty challenges us and often upsets us. Our natural desire to have the world a predictable place and to be in control of situations as they unfold can lead to dysfunctional responses to uncertainty. Organizations are experiments in progress. We are in the process of finding out what does and does not work. Wisdom is an essential tool to have in the face of uncertainty and wisdom is an attitude rather than a skill or a body of knowledge.

Understanding organizations as CAS demonstrates the need for a basic rethinking of the notion of uncertainty. The purpose of this conference was to contribute to this rethinking. The chapters that follow present the thinking of a number of scholars and practitioners about the nature of fundamental uncertainty.

The first section presents the foundation of uncertainty and surprise. In addition to this Chapter, which is an introduction and provides a background on the main themes of complexity science, this section includes a commentary by Professor Prigogine. In Chapter 2, Prigogine challenges the misconception that entropy is associated with disorder by showing that non-equilibrium steady states are characterized by a minimum of entropy production.

The second section focuses primarily on the central themes of uncertainty and surprise. There is an integrative summary of the conference, Chapter 3, by Dean Driebe and Reuben McDaniel. In Chapter 4, Peter Allen's keynote address describes how uncertainty and surprise relate to the evolution of complex systems. Karl Weick's address, in Chapter 5, uses the West Nile Virus incident to illustrate how to organize and manage an organization in the face of the unexpected.

The third section, which is divided into four subsections, presents the differing views of uncertainty and surprise via presentations given by the diverse group of panelists. Subsection A describes the fundamental unknowability in science and social science. In Chapter 6, Linda Reichl draws conclusions relating to change in complex social systems based on similarities seen in the change behavior of complex chemical systems. Later, in Chapter 7, Scott Kelso discusses coordinated dynamics and its relationship to self-organization and knowledge creation. In Chapter 8, Bruce West uses dimensional analysis to explain how a particular aspect of knowability relates to the modeling and simulation of complex phenomena.

Subsection B explores key organizational issues of uncertainty. Erich Baier, in Chapter 9, provides a description of complexity science in business through the eyes of an engineer. Larry Liebovitch continues this subsection in Chapter 10 by focusing on mathematical based approaches in complexity science and by posing the question, "What is the nature of the relationship between the mathematics and the metaphorical lessons that we learn from it?" Chapter 11 concludes this subsection with an example of uncertainty in healthcare delivery in which James Begun asks the question "How would healthcare delivery be different if uncertainty were widely recognized, accurately diagnosed, and appropriately managed?"

Subsection C primarily deals with uncertainty in healthcare delivery. Benjamin Crabtree, in Chapter 12, focuses his thoughts on uncertainty and surprise in the primary care setting. In Chapter 13, John Pierce uses two patient case studies to illustrate the importance of asking the right questions in attempting to avoid medical error. The last chapter in this subsection, written by James Taylor, shares his story from the perspective of the CEO of a teaching hospital and discusses the importance of organization and leadership in hospitals.

Subsection D presents fundamental uncertainty in business and business decision making. In Chapter 15, James Dyer focuses on the manner in which uncertainty is currently evaluated in business, with an emphasis on economic measures. June Holley continues this theme of fundamental uncertainty in business in Chapter 16 by describing how to effectively apply complexity science and networking theory toward transforming your regional economy. Finally, Tom Petzinger concludes this subsection in Chapter 17 by arguing that since the evolution of business is a given, businesses should design systems in which "uncertainty is treated as a certainty."

The fourth section captures some of the informal remarks made during question and answer sessions and small group summary sessions. Hopefully this material will inspire new thinking about the nature of uncertainty and surprise.

References:

Anderson RA, McDaniel RR Jr (1999) RN participation in organizational decision making and improvements in resident outcomes. Health Care Management Review 24(1):7-16

Ashmos DP, Huonker JW, McDaniel RR Jr (1998) Participation as a complicating mechanism: The effect of clinical professional and middle manager participation in hospital performance. Health Care Management Review 23(4):7-20

Begun JW (1985) Managing with professionals in a changing health care environment. Medical Care Review 42(1):3-10

Bossomaier T, Green D (1998) Patterns in the sand. Perseus Books, Reading

Callen E, Shapero D (1974) A theory of social imitation. Physics Today, July

Capra F (1996) The web of life. Anchor Books, New York

Casti JL (1997) Would-be worlds - How simulation is changing the frontiers of science. Wiley, New York

Cilliers P (1998) Complexity and postmodernism. Routledge, New York

Coleman HJ Jr (1999) What enables self-organizing behavior in businesses. Emergence 1(1):33-48
Crabtree BF, Miller WL, Aita V, Flock SA, Stange KC (1998) Primary care practice organization and preventive services delivery: A qualitative analysis. The Journal of Family Practice 46(5):403-409
Dutton J, Jackson SE (1987) Categorizing strategic issues: Links to organizational actions. Academy of Management Review 1:76-90
Eisenhardt KM (1990) Speed and strategic choice: How managers accelerate decision making. California Management Review 2:39-54
Fontana W, Ballati S (1999) Complexity - Why the sudden fuss? Complexity 3(3)
Goldberger AL (1997) Fractal variability versus pathologic periodicity: Complexity loss and stereotypy in disease. Perspectives in Biology and Medicine 40(4):543-561
Goldstein J (1999) Emergence as a construct: History and issues. Emergence-Journal of Complexity Issues in Organizations and Management 1(1):49-72
Holland JH (1998) Emergence - from chaos to order. Addison-Wesley, Reading
Holland JH (1995) Hidden order. Addison-Wesley, Reading
Johnson S (1995) Emergence. Scribner, New York
Kauffmann S (1993) The origins of order. Oxford University Press, New York
Kauffmann S (1995) At home in the universe. Oxford University Press, New York
Izard C (1977) Human emotions. Plenum, New York
Lazarus R (1991) Emotion and adaptation. Oxford University Press, New York
Mainzer K (1996) Thinking in complexity. (2nd ed) Springer, New York
Mandler G (1975) Mind and emotion. Wiley, New York
McDaniel RR Jr, Jordan ME, Fleeman BF (2003) Surprise, surprise, surprise! A complexity science view of the unexpected. Health Care Management Review 28(3):266-278
McKelvey B (1999) Avoiding complexity catastrophe in co evolutionary pockets: Strategies for rugged landscapes. Organization Science 10(3):294-321
Newman DV (1996) Emergence and strange attractors. Philosophy of Science 63(2):245-261
Nicolis G, Prigogine I (1989) Exploring complexity. Freeman, San Francisco
Perrow C (1984) Normal accidents: Living with high-risk technologies. Basic Books, New York
Plutchik R (1984) Emotions: A general psychoevolutionary theory. In: Scherer KR and Ekman P (eds) Approaches to emotion. Erlbaum, Hillsdale, pp 197-219
Roseman IJ (1984) Cognitive determinants of emotion: A structural theory. Review of Personality and Social Psychology 5:11-36
Simon HA (1991) Bounded rationality and organizational learning. Organization Science 2(1):125-134
Waldrop MM (1992) Complexity - the emerging science at the edge of order and chaos. Touchstone/Simon and Schuster, New York
Weick KE, Sutcliffe KM (2001) Managing the unexpected. Josey-Bass, San Francisco
Yourstone SA, Smith HL (2002) Managing system errors and failures in health care organizations: Suggestions for practice and research. Health Care Management Review 27(1):50-61

2 Surprises in a Half Century

Ilya Prigogine[1]

Ilya Prigogine participated early in the organization of our meeting and was invited to give a keynote address. Due to poor health he was unable to attend and sent the following remarks, which were read by his close collaborator Gonzalo Ordonez. Dr. Ordonez has written the footnotes to the remarks and the text has been edited by him and Dean Driebe for inclusion in this volume. We deeply appreciate this contribution of the late Professor Prigogine and profoundly regret his passing, which occurred on May 28, 2003.

Dear friends,

I regret not being able to participate in this interesting workshop. I find this subject quite intriguing. It is true that the development of theoretical physics, especially in my fields of thermodynamics and complex systems, has led to many surprises. It is now sixty years that I have been involved in these problems. I would like to mention a few of these surprises, which occurred over this long period.

My first paper on non-equilibrium physics appeared in 1945. Most people associate entropy with disorder but I have shown that non-equilibrium steady states are characterized by a minimum of entropy production. Entropy changes in a system due to internal entropy production and entropy flow from the environment. The formulation of the second law of thermodynamics is that the entropy production is positive except at equilibrium when it vanishes.

In a constrained system, entropy production takes its minimum value. The entropy itself is generally lower than the equilibrium entropy of the unconstrained system. Let us consider an example of a box separated by a membrane into two parts. It contains a mixture of gases, hydrogen and nitrogen, for example. At equilibrium the concentration of hydrogen and nitrogen will be equal in both parts, but if you heat one boundary the concentrations will change. In one part the concentration of nitrogen will increase and in the other part the concentration of hydrogen will increase. This lowers the entropy. This theorem of minimum entropy production is valid near equilibrium. In the "linear regime," constraints lead to a structure.

This was my first surprise, but as I mentioned the minimum of entropy production was only true in the linear regime. With the help of my colleague Paul Glansdorff we have studied what happens far from equilibrium in the domain of "nonlinearity." We have been most surprised that generally we see the emergence of new structures. The new structures arise at bifurcation points. Bifurcation points

[1] Regental Professor and Ashbel Smith Professor of Physics and Chemical Engineering and Director, Centre for Studies in Statistical Mechanics and Complex Systems, The University of Texas at Austin, and Director, International Solvay Institutes for Physics and Chemistry, Brussels, Belgium (deceased).

correspond to a change in stability. The solution below the bifurcation point becomes unstable and new stable solutions emerge. Bifurcation points exist certainly in social or economic systems. This requires autocatalytic processes.

Curiously, far from equilibrium, the behavior of the system is opposite to the behavior near equilibrium. Near equilibrium the system tries to come as close as possible to vanishing entropy production. Far from equilibrium nature invents new mechanisms that taken separately would permit to the system to approach equilibrium more rapidly. The classic example is Benard instability. This corresponds to a fluid layer heated from below. Near equilibrium we have thermal conduction. Far from equilibrium we have in addition convection and the heat flow increases. We called such structures dissipative structures. They correspond to a kind of self-organization.[2]

These ideas have been developed enormously over the last decades. We see dissipative structures everywhere. The classical difference is between a crystal and a town. A crystal is an equilibrium structure. A town is only possible by the exchanges with the outside world. It is in a sense a dissipative structure.

A surprise for us is that very simple evolution equations may describe highly complex dissipative structures. This surprising result has been rediscovered again and again in the last decades. Dissipative structures are generally connected with symmetry breaking. Consider a system that contains two centers; for example, two shopping centers, with two paths to go from one center to the other. Near equilibrium we have "equi-partition." Both paths are equally populated. Far from equilibrium one path may be nearly full and the other nearly empty. My colleague and friend Jean-Louis Deneubourg has verified that this is exactly what happens in insect societies. There are two nests connected by two bridges. If the insect society is small both bridges show the same population density. If the ant society becomes larger, there is a symmetry breaking: one bridge carries the majority of the ants.

I come now to another class of surprises. It is now 130 years ago that Boltzmann founded kinetic theory and in his approach he was able to introduce entropy

[2] One of the simplest examples of a complex system can be observed when you boil water. As you increase the temperature of the water up to a certain critical temperature, the motion of the molecules is very disorganized. All of the molecules are evenly distributed and going in different directions. But after the temperature increases beyond a certain point, you start to see a rolling motion of the molecules going up and down; this is called convection. This is a common phenomenon in nature and is responsible for much of the behavior of atmospheric and water currents that affect our weather. This is a simple example of a dissipative structure in a complex system. It is a complex system because the water molecules can collectively display a variety of behaviors. Another important point in this example is symmetry breaking. Before reaching the critical temperature, we had evenly distributed molecules moving in a disorganized fashion. As the system crosses the critical temperature, it begins to rotate. But the molecules could rotate clockwise or counter-clockwise. So at some point, the system of molecules makes a choice. This is a bifurcation point, a phenomenon that can also be observed in social systems such as human history. Finally, this example also demonstrates the existence of "long range correlations." While molecules only interact with a handful of their immediate neighbors, in the convection of boiling water, we see the organized motion of huge numbers of molecules.

associated with the velocity distribution of particles.[3] But his results were sharply criticized. The entropy increases in the future. Time symmetry is broken. How it is possible asked Poincaré, to deduce a broken time symmetry starting from time symmetric classical mechanics. The traditional view was that Boltzmann's equation is due to approximations and these approximations introduce irreversibility through some type of coarse graining. This is however very unlikely. We are in the front of a narrative universe. Everywhere in biology, in sociology, in cosmology we see narrative elements. It would be a kind of megalomania that time-directed evolution would be driven by human approximations.

Our group has therefore followed a quite different road. The systems to which Boltzmann's equation apply are a different class of dynamical systems. Every student of physics knows resonance phenomena such as the emission of a photon when the atom goes from an unstable state to the ground state. The systems in which such transitions are essential have a different dynamical behavior. I cannot go into details but only mention that these systems belong to a class that Poincaré called non-integrable systems. We have shown that such systems present time symmetry breaking. Therefore the contradictions in Boltzmann are due not to approximations in the dynamics but to a misuse of classical mechanics. Boltzmann-type equations are correct but we have to enlarge the basic dynamical description in order to understand how they arise. It is one of the surprises of my life that after much research my colleagues and I were able to formulate the basic laws for this enlarged dynamics.

It is clear that these considerations can be applied to social and economic problems. The literature is enormous. I would simply quote a recent paper by Madeleine Beekman, David J. T. Sumpter, and Francis L. W. Ratnieks, "Phase transition between disordered and ordered foraging in Pharaoh's ants." I reproduce the summary of their paper:

> The complex collective behavior seen in many insect societies strongly suggests that a minimum number of workers are required for these societies to function effectively. Here we investigated the transition between disordered and ordered foraging in the Pharaoh's ant. We show that small colonies forage in a disorganized manner, with a transition to organized pheromone-based foraging in larger colonies. We also show that when food

[3] Here we are talking about a gas system, and what is called the Boltzmann equation, named for the Austrian physicist who first formulated it in the 1800's. This is one of the more important equations in Physics because it is the root of a lot of explanations about a variety of phenomenon. Basically, this equation concerns counting how many particles have a certain velocity "V." There are two kinds of processes here; one is loss, the other is a gain. For example, if a given particle collides with another particle, it now has a velocity of V' and we have lost a particle with velocity V. But we could also have the opposite situation, where a particle with a velocity of V' has a collision which changes its velocity to V, resulting in a gain. The gain and loss processes tend to increase the disorder in the distribution of velocities. After some time the system will reach a balance between gain and loss, where nothing will change much anymore. This is when we reach equilibrium. In contrast to Newton's equation, Boltzmann's equation breaks time symmetry.

sources are difficult to locate through independent searching, this transition is first-order and exhibits histeresis, comparable to a first-order phase transition found in many physical systems. To our knowledge, this is the first experimental evidence of a behavioral phase transition between maladaptive (disorganized) and an adaptive (organized) state. (Proceedings of the National Academy of Sciences 2001, 98(17):9703-9706).

In brief, this means that ants self-organize to feed. This is an example of a non-equilibrium phase transition. The analogy with human society is obvious. Human history may be described by a succession of bifurcations. Probably, the first was the transition from Paleolithic to Neolithic society.

Let me quote a final example. My colleague Deneubourg has performed the following experiment. You have two large boxes connected by a small channel. The ant's nest is displaced from one box to the other. Ants are crossing the channel to recover the initial situation. Deneubourg has found a way of identifying individual ants. The surprising result is that there are two groups of ants: active ants and lazy ants. The active ants cross the channel repeatedly. The lazy ants remain most of the time in the same box. Deneubourg selected out the lazy ants and made the same experiment. The astonishing result is that some of the lazy ants then begin to work. How human this sounds!

The starting point of research is, as emphasized by Einstein and others, astonishment. As long as a problem remains unsolved we imagine many solutions. Finally, if you are happy with one of them you support it. The emergence of this solution is often a surprise.

We live in a probabilistic universe. Our universe is one of many possible universes. This universe is an evolving one. The future is not given and therefore we have only a probabilistic description and there is no certainty. Surprise is part of our fate as a part of nature in which we are embedded. We have to accept that we are only a very small part of this evolving universe. Creativity of nature and human creativity cannot be separated. But creativity implies surprises: David of Michelangelo or relativity theory. Uncertainty and surprise are part of human destiny.

Section II Central Themes on Uncertainty and Surprise

3 Complexity, Uncertainty and Surprise: An Integrated View

Dean J. Driebe[1] and Reuben R. McDaniel, Jr.[2]

3.1 Introduction

The purpose of this paper is to tie together some of the major themes that emerged at the conference and in the presentations in this volume. As befits the interdisciplinary nature of the conference, one of the authors of this contribution comes from a hard sciences background in physics and the other is from business management and organization theory. We will give our perspectives on the fundamental science and organization theory behind the themes of the meeting. We will also attempt to capture some of the key ideas that are presented in the papers in this volume. We will then propose a research agenda that addresses some of the major open questions that have become clearer from the work done by the authors of these papers.

Our scientific worldview has changed in the last few decades through the realization that complex phenomena require new tools for their description and engender new theoretical insights. Complexity science has taught us to expect the unexpected in a wide variety of physical, biological and social systems. This changes our basic attitude to the appearance of surprise in the world and allows us to cope with it in new ways. The papers in this volume illuminate many aspects of the changing worldview and identify many strategies for coping.

Surprise is corollary to one's expectations, which depend on what we know about the dynamics of a given system and what information we have on the state of the system (Prigogine, this volume; Reichl, this volume). Surprise is also a corollary of how we make sense of the world we encounter (Weick, this volume; Pierce, this volume; Begun and Kaissi, this volume). Complexity science has demonstrated that there is more to these factors than was previously appreciated. For some systems, acquiring more information about their state allows one to significantly reduce uncertainty about their future states. For others, one doesn't gain much from additional information about the state of the system because the system dynamics itself produces information over time. This has very important practical

[1] Research Scientist, The Ilya Prigogine Center for Studies in Statistical Mechanics and Complex Systems, The University of Texas at Austin, dean@physics.utexas.edu.

[2] Charles and Elizabeth Prothro Regents Chair, McCombs School of Business, The University of Texas at Austin, reuben.mcdaniel@mccombs.utexas.edu.

consequences when, for example, a judgment has to be made to allocate resources to either learn about the state of a system or its underlying dynamics.

In science, as in art or literature, to gain insight into the real world we make an artificial world, which serves as a model that we can manipulate and probe. As is highlighted by West in this volume, the complexity science viewpoint alters our traditional modeling techniques. A complex system challenges an easy separation of relevant from irrelevant variables by presenting situations where so many variables may be relevant that the only modeling possible may be a reproduction of the system itself. If the artificial world we use as our model has to be made essentially identical to the real one, the traditional modeling process loses its power. Complexity science, coupled with technological developments that have given widespread access to computing power, has seen the development of new modeling techniques in order to gain insights into the real world of fundamental uncertainty. Kelso (this volume), Liebovitch (this volume), and Allen (this volume) demonstrate how new modeling techniques can open up new lines of thinking about complex systems.

As noted by Dyer (this volume) the presence of fundamental uncertainty has led to the use of the theory of real options in making economic decisions in business situations to make more sense of uncertain situations and to assist in making reasonable decisions about resource allocation. Baier (this volume) calls our attention to the fact that "one of the major challenges we have in our business (technology development) is the balance between stability, predictability and adaptability to change." The issues surrounding understanding and coping with the uncertainty created by systems dynamics is the central topic of many of the papers in this volume.

The property that puts us in this situation of fundamental uncertainty is nonlinearity. The ramifications of the nonlinearities inherent in so many systems have only recently been widely appreciated and studied. The basic difficulty was well expressed over seventy years ago by the British physicist Sir Arthur Eddington: "We often think that when we have completed our study of one, we know all about two, because 'two' is 'one and one.' We forget that we still have to make a study of 'and'" (Eddington 1928:103-104). One can say that complexity science deals directly with this study of "and." Yet, even the notion of "and" can be challenged in the complexity sciences as noted by Kelso (this volume) who calls our attention to the fact that we often partition the world into pairs when it is the complementarities that offer a window of opportunity for understanding. West (this volume) focuses on evolutionary constructs from complexity science to bring some of the qualitative features of nonlinear systems dynamics to light.

3.2 Complex Dynamics in Nature, Organizations and Society

The contemporary enchantment with complex systems studies was foreshadowed by the mid-20th century work of Ludwig von Bertalanffy and others in *General*

Systems Theory (von Bertalanffy 1968). Ilya Prigogine was also a pioneer in his appreciation of far-from-equilibrium thermodynamics and self-organizing systems leading to his theory of dissipative structures (Prigogine and Stengers 1984). Complexity science emerged as a recognized field in the 1970's. In the following decade its status was heightened by the opening of numerous research centers around the world devoted to aspects of complex system dynamics ranging from so-called artificial life, to the study of chaos in financial markets, to human brain dynamics. A bestselling book in 1987, James Gleick's *Chaos: The Making of a New Science*, brought some of the developments in the field to a mass audience. The Santa Fe Institute is one of the best-known research centers devoted to the study of complexity. Waldrop, in his book *Complexity: The Emerging Science at the Edge of Order and Chaos* (1992), brought the work of the Santa Fe Institute to the view of many members of the public and to members of the scientific community who were unaware of the work being done in this intellectual area. There continues to be a stream of literature, some of it primarily directed at a lay audience, which explicates aspects of complexity science.

A few years ago an issue of the journal *Science* was devoted to complexity science. In a key article, after assessing the main aspects of the discipline, the authors address what we have learned from complexity science: "Up to now physicists looked for fundamental laws true for all times and all places. But each complex system is different; apparently there are no general laws for complexity. Instead one must reach for 'lessons' that might, with insight and understanding, be learned in one system and applied to another. Maybe physics studies will become more like human experience" (Goldenfeld and Kadanoff 1999:89).

Throughout the 90's and into the new century, progress on many fronts has been made but it is fair to say that some of the hype associated with the field a few decades ago has not been realized. It has been especially difficult to make concrete applications of some of the physical and mathematical aspects that have been well developed. It is easy to talk about new perspectives and new insights but it is hard to use them in practice. One of the reasons for this is the difficulty of making generalizations in complex systems studies. Several of the works in this volume indicate some of the approaches that are being made to deal with this concern. For example, Allen provides new insights into the irreversible nature of some decisions in an uncertain world and Crabtree indicates how ideas from complexity are used to understand a set of organizations in new ways. The papers in this volume demonstrate that there have been both analytical and metaphorical uses of ideas from complexity science to further our understanding of uncertainty and surprise.

These new perspectives enable communication and cross-fertilization between the physical sciences and humanistic concerns. Petzinger (this volume) and Holley (this volume) show how they have used ideas from complexity science to help them understand the dynamics of entrepreneurship. Crabtree (this volume) brings complexity science to primary care medical practices to help explain phenomenon that are difficult to understand with conventional organizational theories. Pierce (this volume) suggests some ways that thinking about complex adaptive systems can inform our understanding of medical errors.

Thinking in complexity reaches back at least as far as the pre-Socratics (Mainzer 1994). But progress in science, especially during and since the Enlightenment, has been associated with abstracting the regularities observed in nature to the status of "laws of nature." Complexity science deals with irregularities, with results that are difficult to reproduce and with situations that challenge the "doctrine of the uniformity of the material universe" (Whitehead 1961:7). In this sense complexity science is science discovering its own borders (Berge et al. 1986). There are two broad aspects of complex systems studies that are important to distinguish from each other. These two aspects challenge our intuitive conception that complex behavior always arises in a complex system and that the appearance of simple behavior tells us that the system presenting it is also simple. One aspect is where very simple systems display unexpectedly complex behavior. Such situations, which are typically characterized by sensitivity to initial conditions, are often studied in the context of chaotic dynamical systems and will be discussed in the next section.

The second aspect is the emergence of remarkably coherent phenomena in systems of many interacting elements and this has been studied in the context of self-organization, dissipative structures and complex adaptive systems. In typical situations the elements, which are sometimes called agents, interact nonlinearly and are subjected to far-from-equilibrium conditions. Social examples include the development of informal organizations in companies and the co-evolutional properties one observes in industries. Weick (this volume) and Pierce (this volume) both bring serious social phenomena into much clearer focus using ideas from complexity science. Physical examples include nonequilibrium fluid dynamics and driven chemical reactions, as discussed by Reichl in this volume. In nature and society one can mention the self-organizing behavior of ant colonies and the adaptability of economies. Crabtree (this volume) and Pierce (this volume) indicate how self-organizing behavior manifests itself in health care systems and Holley (this volume) and Petzinger (this volume) show how this plays out in entrepreneurial settings. Allen (this volume) develops a deeper understanding of social economic systems using notions of uncertainty and surprise.

The basic fact that uncertainty and surprise occur in human systems is easily comprehended. But we can gain insight into the workings of social systems - and how they can thus be improved - by borrowing tools and insights employed for complex physical, chemical and biological systems. Holley (this volume) indicates how transformation of a regional economy might be informed by these ideas. And Crabtree (this volume) provides significant evidence that understanding primary care practices in medicine is greatly enhanced through understandings from complexity science. Many of the authors in this volume point to ways in which complexity science thinking can help in unraveling some of the mystery of human systems.

A key insight, elucidated principally by Ilya Prigogine and his groups working in Brussels and Austin, is the connection between macroscopic order and microscopic fluctuations. As Prigogine puts it:

> In most of the phenomena studied in classical physics, fluctuations play only a minor role. This is the case in the whole domain of classical equilibrium thermodynamics based on Boltzmann's ordering principle. On the other hand, the study of nonlinear systems under conditions far from equilibrium leads to new situations in which fluctuations play a central role. It is the fluctuations that can force the system to leave a given macroscopic state and lead it on to a new state which has a different spatiotemporal structure. (Prigogine 1976:93)

This new appreciation for the crucial role of fluctuations in system dynamics runs through many of the presentations in this volume. See, for example, Pierce's effort to apply these notions to the pressing issue of medical errors and Allen's description of symmetry breaking in social systems. Since microscopic fluctuations are fundamentally unknowable this enables us to comprehend the fundamental unknowability of the type of macroscopic order a system may take. Conversely, it may enable a leveraging in certain situations, such as social ones, of a controllable microscopic dynamics to influence widespread macro behavior. For example, Begun and Kaissi (this volume) discusses the role of professional actors in absorbing societal uncertainties.

Nearly any activity involving human interaction is permeated with nonlinearities. Many papers in the volume explore the human aspects of complexity, uncertainty and surprise. There are elements present in human systems that are nearly completely foreign to physical systems, namely memory and anticipation. Memory effects, such as hysteresis in magnetic systems, are present in some physical systems, but anticipation appears to be a hallmark of intelligence. One of the key elements of mind is the ability to see through time, both behind and ahead. This ability is limited though, due in part to issues addressed throughout this volume.

Weick (this volume) suggests that some complexity in human endeavors can be a function of the distributed sensemaking capabilities of social systems. He goes on to indicate how some concepts from cognition, sensemaking, workflow interdependence and interrelating can help in providing insights into the ways that complexity analysis fits human organizations. The human condition adds to complexity but the complexity of humans serves to enable effective action in complex situations. When we recognize that at some level, uncertainty and surprise are socially constructed phenomenon (Begun and Kaissi, this volume) we add to the difficulty we have in understanding uncertainty and the role that memory and anticipation play in the existence of uncertainty and in the capacity of systems to cope with uncertainty.

There are also intent and emotion at play in human systems. In a hospital crisis, as noted by Taylor (this volume), this intent and emotion often becomes central in understanding and managing the crisis. The processing of all these factors by the agents in the systems and the subsequent response then changes the dynamics of the system itself. Taylor (this volume) goes on to suggest that his response to an organizational crisis was grounded in his understanding that the hospital was not a machine with simple dynamics but a system with complex dynamics. This response was out of step with conventional management wisdom. In social systems knowledge affects reality and changes the facts. Begun and Kaissi (this volume) alert us to the importance of the social construction of uncertainty in health care and the role of clinical professionals in the "management" of that uncertainty.

They indicate that acceptance of fundamental uncertainty, as difficult as it might be, could lead to different, and perhaps more reasonable, health care strategies and policies. Weick (this volume) shows how a system's intent can be undermined by its sensemaking capacity. Alterations in the interrelationships in a system can lead to the reorganization of sensemaking capacity and open up new opportunities for achieving an organization's goals.

Among the insights that emerge from the intersection of the papers in this volume is the critical role that humans and human nature play in dealing with uncertainty and surprise. Another is the role of organizations as complex adaptive systems as both generators and absorbers of surprise. Several new approaches to modeling complex systems and their dynamics are suggested. Perhaps one of the clearest insights is the role that complexity science plays in providing concepts and constructs that are used metaphorically to increase our understanding of social systems and their dynamics. The papers in this volume come at these concerns from a variety of perspectives and this is appropriate given the strong interdisciplinary focus of complexity science.

3.3 Chaotic Dynamics, Uncertainty and Surprise

Larry Liebovitch in this volume tells us that "not everything that looks random is random." He is talking about deterministic chaos. This is random-looking behavior that arises in a system, typically of very few (even one is enough!) variables with completely specified, i.e., deterministic evolution rules. That complex behavior can occur in a simple system has been known since the late 19th century when the French mathematical physicist Henri Poincaré revealed that this was happening in the gravitational three-body problem. It is only in the last few decades of the 20th century that the phenomenon of chaos that Poincaré discovered has taken off as a "new science" (Gleick 1987).

The modern realization of the widespread occurrence of chaotic dynamics is generally ascribed to Edward Lorenz, a meteorologist who serendipitously noticed sensitivity to initial conditions in a computer simulation of a weather model (Lorenz 1993). In the late 1950's Lorenz was constructing model systems of equations in order to test competing weather prediction schemes. He settled on a twelve variable model that displayed "interesting" non-periodic behavior. As Lorenz tells it in his book:

> At one point I decided to repeat some of the computations in order to examine what was happening in greater detail. I stopped the computer, typed in a line of numbers that it had printed out a while earlier, and set it running again. I went down the hall for a cup of coffee and returned after about an hour, during which time the computer had simulated about two months of weather. The numbers being printed were nothing like the old ones. I immediately suspected a weak vacuum tube or some other computer trouble, which was not uncommon, but before calling for service I decided to see just where the mistake had occurred, knowing that this could speed up the servicing process. Instead of a sudden break, I found that the new values at first repeated the old ones, but soon afterward differed by one

and then several units in the last decimal place, and then began to differ in the next to the last place and then in the place before that. In fact, the differences more or less steadily doubled in size every four days or so, until all resemblance with the original output disappeared in the second month. This was enough to tell me what had happened: the numbers that I had typed in were not the exact original numbers, but were the rounded-off values that had appeared in the original printout. The initial round-off errors were the culprits; they were steadily amplifying until they dominated the solution. In today's terminology, there was chaos. (1993:134-136)

This amplification of errors is one of the characteristic features of chaotic dynamics. In contrast, a system with non-chaotic or regular dynamics will not display such sensitivity to initial conditions. This fundamental distinction in the types of dynamics a closed autonomous system may display has only recently gained appreciation. It was certainly realized by Poincaré and others following him concerned with dynamical systems but even though it was brought to a wider audience in the physical sciences since the 1970's it has as yet not seeped into general public consciousness. Our intuition tells us that random behavior occurs because a systems is subject to many influences that escape our detection. This is surely the case in many situations but chaotic dynamics is something else.

A striking illustration of deterministic chaos is furnished by the behavior of a double pendulum. This simple mechanical device consists of a pendulum composed of a stiff rod, hung to enable it to rotate freely, with another rod attached at its end also able to rotate. For small excursions from their resting places the two pendula behave as two coupled simple harmonic (linear) oscillators and exhibit regular periodic behavior. Even though there are collective modes present in the composite system, the behavior of each member is still just simple oscillatory motion. For large excursions resulting from a big push on the system there occurs a completely different type of behavior. The motion of the two pendula is apparently random as though the system is driven by complex external forces but it is entirely autonomous and due just to the dynamics of a two-component simple system. For large excursions the two pendula are now nonlinear oscillators. Resonance between the two oscillators is occurring as Reichl describes in this volume. This is a striking example of deterministic chaos - here occurring in a system governed by classical Newtonian dynamics.

Coupled nonlinear oscillators like the double pendulum are notoriously difficult to study in full detail. There is no solution to the equations of motion because such systems are nonintegrable. This is a property that Poincaré had discovered in his study of the three-body problem. Integrable systems, like the two-body gravitational problem or the simple harmonic oscillator have constraints (so-called integrals of motion) that keep their dynamics confined to smooth surfaces in phase space and enable solutions in terms of smooth regular functions to be written down - like the sinusoidal solution for harmonic motion. In nonintegrable systems these constraints are lifted and the system is free to explore more regions of the phase space while still being consistent with energy conservation.

There is another fundamental aspect that distinguishes regular from chaotic and nonintegrable dynamics that is relevant to the contents of this volume. It concerns an information-theoretic viewpoint of the dynamics. In general we have some dynamical rule that enables us to write down equations of motion for the system. We need to then have initial conditions of the dynamical variables in order to calculate future values. In a deterministic system the initial conditions along with the equations of motion enable all future (or past!) states to be determined. But this determination, and its dependence on the initial conditions is very different depending on whether the system is regular or chaotic.

Generally speaking, in a regular system if we know the initial conditions to some accuracy then we can use the equations of motion (or their solution) to determine any future state to the same accuracy. If we want to know better what will happen later we find out more about what is happening now. In a chaotic system, knowing the present state to some accuracy only enables us to know about future states for some limited amount of time (the predictability horizon). And our knowledge of the future degrades rapidly until we have used up all the information given in the initial condition. Knowing more about the present enables us to know a little more about the system for a little more time. This is because the dynamics in a sense unfolds the initial condition in time and needs more and more of the initial condition (exponentially in time!) to determine the unfolding of the system.

Because we can't know everything about the present state of a chaotic system we will always be surprised by its future states. We also don't know everything about present states of systems with regular dynamics but they will not surprise us because the amount we know is preserved in time. What we know now about a chaotic system is lost in time. We have fundamental uncertainty.

Physicists have accepted fundamental uncertainty since the 1920's when quantum mechanics was formulated to account for behavior on microscopic levels.[3] But this uncertainty doesn't seem to alter our intuitive conceptions, at least since the enlightenment, of the deterministic nature of the world on our level. Chaos does alter these conceptions and provides a perspective of uncertainty created entirely by the internal dynamics of simple autonomous closed systems - even those governed by deterministic Newtonian dynamics.

It is useful to list some types of uncertainty that are encountered. We start with situations where the uncertainty can be reduced by acquiring information and move up to those where the uncertainty is fundamental and irreducible.

Lack of knowledge of a simple process. The uncertainty here is eliminated once the process is known - or over time as it is observed.

Reduced dynamics of an open system. In this situation we are only aware of the dynamics of part of a system. Often such cases are analyzed as a system coupled to an environment that introduces unforeseen "shocks" to the system of interest. If the environmental dynamics can be discerned then the uncertainty may be eliminated.

---

[3] Holdouts to the uncertain world presented by quantum mechanics have clung to a view that the probabilistic aspects of the theory mask an underlying determinism. The most well-known proponent of this viewpoint was Einstein, who famously declared that "God does not play dice." Over seventy five years of search for such a theory by fringe groups has so far been deemed unsuccessful.

Chaotic dynamics. This situation is characterized by exponential sensitivity to initial conditions on the level of trajectories. Knowing more about the initial conditions is of very limited use for predicting the future time development of a trajectory.

Irreducible complex dynamics with many degrees of freedom. This is the situation that is most challenging in a classical dynamics context. It is encountered with fluid turbulence and the weather. There may be chaotic aspects to the dynamics but these are in general difficult to disentangle from the complications introduced by the many elements of the system.

Reflexive dynamics. This occurs when the system is composed of thinking agents who react to the development of the system by changing the way they think about it thus changing the dynamics. This situation has been extensively discussed by Soros in the context of economic systems (Soros 1987).

Quantum dynamics. Here only a probabilistic description is possible.

It should be noted that these are all considered in one way or another in the papers in this volume. From the level of chaotic dynamics and above we have fundamental uncertainty in that surprises in the development of the system can never be eliminated. To deal with such situations probabilistic forecasts become increasingly necessary. Dealing with uncertainty in real world situations using probabilistic thinking is nicely discussed in a book by former U.S. Treasury Secretary, Robert E. Rubin (Rubin and Weisberg 2003).

3.4 Toward a Research Agenda

We have seen that the complexity viewpoint gives us new categories of thought. It also provides new techniques for the analysis of system dynamics. The papers in this volume give testament to the broad range of perspectives opened up by viewing systems, whether human or physical, in terms of complexity and chaos. But there are a plethora of important unanswered questions that emerge from the considerations in this volume. Some that can be posed are:

What are the responsibilities of workers and managers when they live in an uncertain world? Do their roles change and how do they change? How can workers and mangers be held accountable for outcomes in an uncertain world, or should they? Should organizations develop performance measures that seek better knowledge, not better results?

How is the work life of managers and workers affected by the fundamental uncertainty that surrounds them? What can be done in schools and other societal institutions to help people cope with the anxieties associated with living in a fundamentally uncertain world?

There is often the need for sensemaking because of uncertainty in the environment rather than because of weaknesses in agents. What can we do so that workers and managers are better at sensemaking in a world that is fundamentally uncertain? What strategies can be developed to incorporate this sensemaking more fully into decisions and into programs for action?

Some have suggested that ideas of complexity are useful as metaphors even in contexts where the ideas themselves may not seem to be directly applicable. What are the strengths and weaknesses of the use of metaphor in a fundamentally uncertain world? Can metaphors be used to structure complex problems so that they are more amenable to analysis? Can guidelines be developed so that the use of such metaphors is truly beneficial?

In an open system coupled to an environment, such as a business organization, how can one distinguish surprise arising from intrinsic system dynamics, such as chaotic dynamics, surprise arising from emergence in a self-organizing system or surprise due to unknown outside influences? What are the relationships between these? Can they be distinguished by an agent embedded in the system and can this be accomplished even with very partial information?

What are the right metrics to use in human systems to enable one to distinguish chaotic dynamics with intrinsic uncertainties from regular dynamics subject to unknown shocks?

Even in an uncertain environment it is necessary for an organization to have stability to function. How can one achieve a balance between the adaptability required to cope with the uncertainty and the stability required to function?

In an organization where there is fundamental uncertainty how can one help people to better understand the interdependencies of the agents and the pivotal role of these interdependencies in the healthy functioning of the organization? There is a strong tendency for people to attempt to decouple agents in order to better understand them and control them. What strategies other than reductionism can people learn to use to manage interdependencies?

When you are confronted with uncertainty and surprise how can you tell if what you are doing is working? How do we account for surprising emergent properties or surprising patterns of self-organization?

In organizations there are significant variations in perceptions of the world. Socialization is a powerful shaper of perceptions and people in organizations are often purposely socialized differently to achieve the diversity required for success (for example, nurses and doctors). Given the fundamental uncertainties that are encountered, what are the consequences of these varying perceptions and can they be reconciled so that progress can be made?

Can the ideas on control of chaos used in dynamical systems theory be transferred to tame human organizations that have chaotic components in their dynamics?

Is it possible to identify macroscopic systems, such as organizations, that have characteristics that can truly be identified as quantum mechanical in nature beyond metaphorical comparisons?

3.5 Conclusions

The late physicist Heinz Pagels has written that "The reason science works is because it studies an ordered world that can be known by an ordered mind"

(1988:244). How Kantian this statement is! But clearly this is only part of the story. Our world also presents disorder in many aspects as well as the surprising appearance of "order out of chaos." These aspects should also be amenable to scientific analysis and have become so in recent decades.

A balanced understanding of our physical world enables a balanced understanding of our biological and social worlds as well. We can cope with uncertainty when we realize that it is not just due to a lack of knowledge of a complicated system but is part of the basic fabric of our universe on all levels - not only the microscopic one described by quantum mechanics.

A paragraph from Pagels' book *The Dreams of Reason* is relevant to our discussion:

> Ultimately, all scientific activity - the thinking, observation, and experimentation - is devoted to the search for a coherent, conceptual representation of reality - a theory or picture of reality. Once one has such a picture of reality or even a partial picture, it has a rich implicative structure, not only for science, but for culture, technology, and commerce as well. Such pictures of reality show us new aspects of the building code of the universe, creating a mental picture that goes beyond anything we can grasp directly with our senses or instruments. And the motivation behind finding these theories is the scientist's desire to know what in heaven and hell is really going on. (1988:161)

Some of these new pictures of reality are presented in the papers in this volume. Many of these pictures show a very healthy intersection of the social and the natural sciences. One of the promises of complexity science has been to focus attention on the way that things are alike, rather than the way they are different. The new pictures of reality that are presented in this volume certainly speak to this promise.

References

Berge P, Yves P, Christian V (1986) Order within chaos. Wiley-Interscience, New York

Eddington A (1928) The nature of the physical world. Macmillan, New York

Gleick J (1987) Chaos: Making a new science. Viking, New York

Goldenfeld N, Kadanoff L (1999) Simple lessons from complexity. Science 284:87-89

Mainzer K (1994) Thinking in complexity: The complex dynamics of matter, mind, and mankind. Springer, New York

Lorenz E (1993) The essence of chaos. University of Washington Press, Seattle

Pagels H (1988) The dreams of reason: The computer and the rise of the sciences of complexity. Simon and Schuster, New York

Prigogine I (1976) Order through fluctuation: Self-organization and social system. In: Jantsch E, Conrad H (eds) Evolution and consciousness: Human systems in transition. Waddington, Addison-Wesley, Reading

Prigogine I, Stengers I (1984) Order out of chaos. Bantam, New York

Soros G (1987) The alchemy of finance. Wiley, New York

Rubin RE, Weisberg J (2003) In an uncertain world. Random House, New York

Von Bertalanffy L (1968) General system theory: Foundations, development, applications. George Braziller, New York

Whitehead AN (1961) The first physical synthesis. In: Johnson, AH (ed) The interpretation of science: Selected essays. Bobbs-Merrill, Indianapolis

4 The Evolutionary Complexity of Social Economic Systems: The Inevitability of Uncertainty and Surprise

Peter M. Allen[1], Mark Strathern[2], and James S. Baldwin[3]

4.1 Introduction

In order to improve our quality of life and the successful functioning of our organisations and social institutions, or to mitigate some anticipated problem, we need to understand how they "work" and to be able to explore the probable consequences of different possible policies or interventions. And this means that we need to "understand" how the underlying causality of a current social situation is operating, and how it might respond to some changed rules or policies aimed at modifying it in a favourable way. Traditionally, however, this would be thought of as understanding the situation as a "mechanical system" with the different actors locked in a predictable system of interaction, that our "policy intervention" would seek to modify in order to improve the overall outcome. The difficulty with this approach however, is that it fails to recognise the essentially fluid nature of human behaviour, and the ability of actors to modify their previous habits in response to the new opportunities or constraints of the situation. In other words, if our policies and interventions are to succeed, we must attempt to anticipate to some degree the different kinds of outcome that we might provoke. This means that we need to move from a mechanical representation to an evolutionary one, in which structural transformations can and do occur.

In several previous papers (Allen 2000, 2001a, 2001b), it was shown how the creative interaction of multiple agents is naturally described by co-evolutionary, complex systems models in which both the agents, the structure of their interactions and the products and services that they exchange evolve *qualitatively*.

[1] Professor of Evolutionary Complex Systems, Head of Complex Systems Management Centre, School of Management, Cranfield University, UK, p.m.allen@cranfield.ac.uk.
[2] Complex Systems Management Centre, Cranfield University, UK.
[3] Advanced Manufacturing Research Centre, Department of Mechanical Engineering, University of Sheffield, UK.

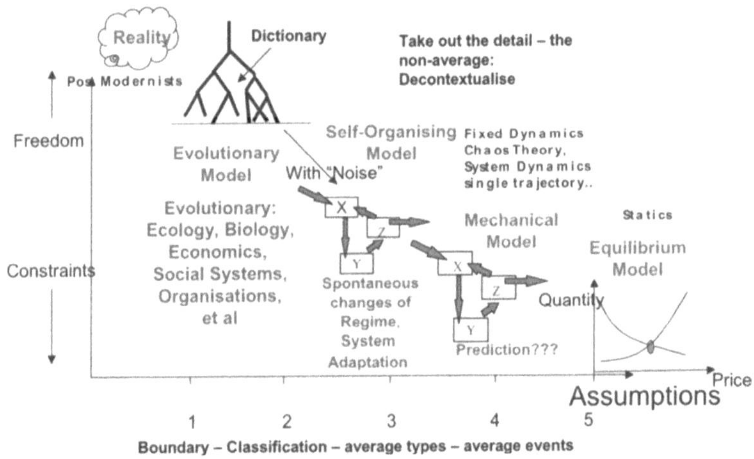

Fig. 4.1. The overall conceptual scheme of increasingly simplified representation of a situation, as increasingly strong assumptions are made

In reality, complex systems thinking offers us a new, integrative paradigm, in which we retain the fact of multiple subjectivities, and of differing perceptions and views, and indeed see this as part of the complexity, and a source of creative interaction and of innovation and change. The underlying paradox is that knowledge of any particular discipline will necessarily imply "a lack of knowledge" of other aspects. But all the different disciplines and domains of "knowledge" will interact through reality - and so actions based on any particular domain of knowledge, although seemingly rational and consistent, will necessarily be inadequate.

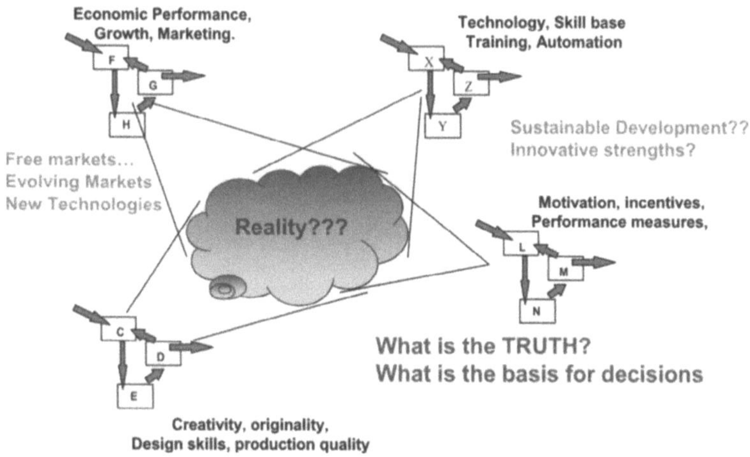

Fig. 4.2. Different people see the same system in different ways. Each can however be rational and consistent, whilst implying different actions or policies

Management, or policy exploration require an integrated view. These new ideas encompass evolutionary processes in general, and apply to the social, cultural, economic, technological, psychological and philosophical aspects of our realities. Often, we restrict our studies to only the "economic" aspects of a situation, with accompanying numbers, but we should not forget that we may be looking at very "lagged" indicators of other phenomena involving people, emotions, relationships, and intuitions - to mention but a few. We may need to be careful in thinking that our views will be useful if they are based on observations and theories that refer only to a small sub-space of reality - the economic zone. The underlying causes and explanations may involve other factors entirely, and the economic "effects" of these may be only delayed ripples or possibly tidal waves. What matters over time is the expansion of any system into new dimensions and conceptual spaces, as a result of successive instabilities involving dimensions additional to those the current "system" appears to occupy.

This idea of evolution as a question of "invadability," with respect to what was not yet in the system, was the subject of a very early paper by the author (Allen 1976). Essentially then, systems are seen as temporary, emergent structures that result from the self-reinforcing non-linear interactions that result from successive "invasions." History is written not only by some process of "rational improvement" in its internal structure but more fundamentally by its dialogue with elements that are *not yet* in the system - successive experimental linkages that either are rejected by the system, or which "take off" and modify the system irreversibly.

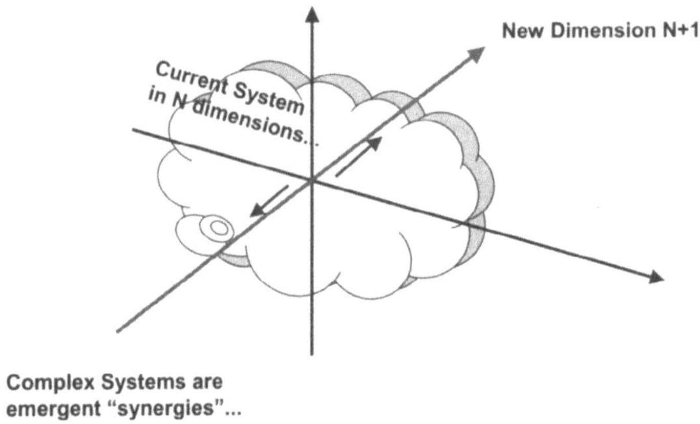

Fig. 4.3. The evolution of complex systems, at different possible levels within structures, is a "dialogue" with the aspects and factors that are not playing an active part within it, at present

Rational improvement of internal structure, the traditional domain of systems' thinking," supposes that the system has a purpose, and known measures of "performance" which can indicate the direction of improvements. But, this more fundamental structural evolution of complex systems that results from successive invasions of the system by new elements and entities is characterized by emergent properties and effects, that lead to new attributes, purposes and performance measures. In the next sections therefore, we attempt to show that this structural evolution is not in fact "random" in its outcome, as successful invasions of a system are always characterized by the revelation of positive feedback and synergy, creating particular new, internally coherent, structures from a growing, explosively rich set of diverse possibilities.

4.2 Mechanical and Self-Organising Representations

Let us consider a very simple social science example of a model of reality that starts from the right-hand side of the Figure 4.1. It will illustrate how non-linear responses can generate new (false) information, can break symmetries and lead on to evolutionary change.

The problem will be that of policemen attempting to catch criminals. In our first model we shall simply assume that policemen move randomly around the town, and stop and search people when they appear to be acting suspiciously. Now, in this model we are going to suppose that there are two types of people present in the city - pink and blue people (as fits the colours of an Excel program). These two types of people commit crime to exactly the same extent (3% crime rate/person/year), and hence policemen acting randomly find just as many criminals that are blue or pink. In our model of this situation, we find a result of 3% exactly.

However, if we consider this a bit more carefully then we find that there is in fact a small sampling error in which pure chance can accidentally create small deviations, according to some normal distribution, around the average rate. We find the result as shown in Figure 4.4.

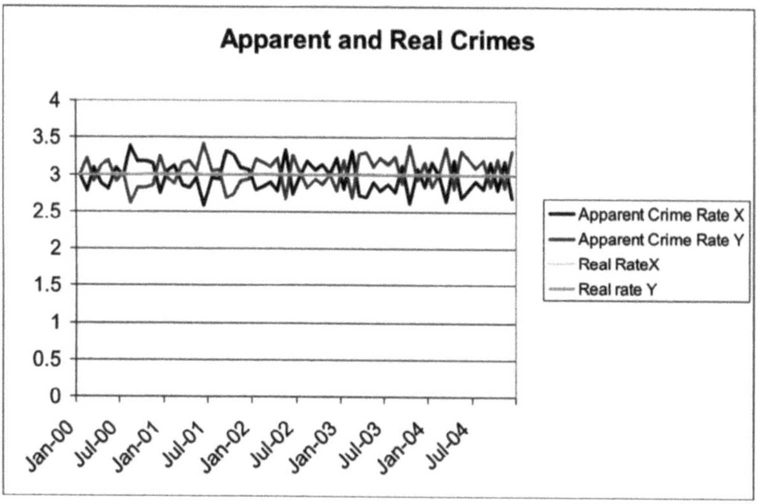

Fig. 4.4. Small fluctuations occur around the average value

Now, this is all very fine, except that policemen cost quite a bit to the community and so it is natural to demand that they perform as effectively as possible, and that they give as good value for money as possible. To achieve this, the police authorities may decide that policemen will be promoted and rewarded according to their success in arresting criminals. This sounds fine, and so in the next case policemen respond to these performance incentives by theorising about who may be committing more crimes, and in targeting those types more than others.

The result varies each time it is run, but essentially, the main point is that a result such as that shown in Figure 4.5 emerges.

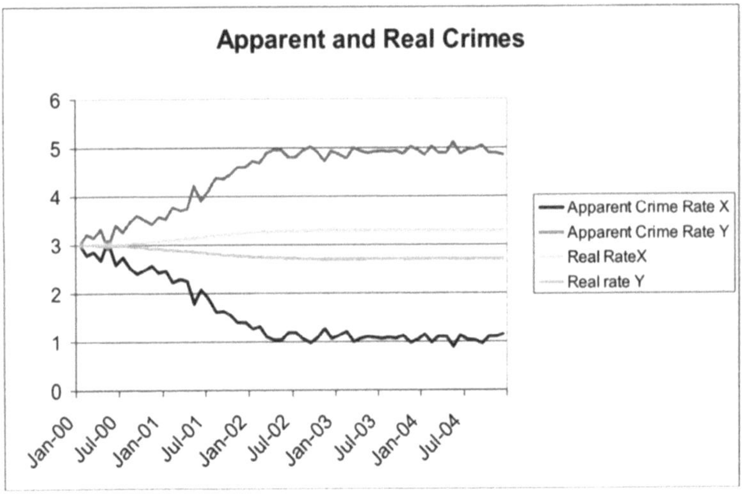

Fig. 4.5. If we turn the targeting parameter up from 1 to 7 then we get an immediate result as the targeting leads either to increased apparent criminality in either pink or blue

So, increased targeting produces a result, but in fact the knowledge that emerges is false, as it still remains true that the crime rate for both types is identical. In fact, in this version, we have even assumed that a population that is hardly ever targeted may start to commit more crimes, since they seem very likely to escape detection, and so the real crime rate of blue is greater than that of pink, but the apparent crime rate of pink is greater than blue. So, what has happened is that policemen have theorised, and have allowed their experience to guide their actions, their actions to generate their results, and their results to confirm their experience.

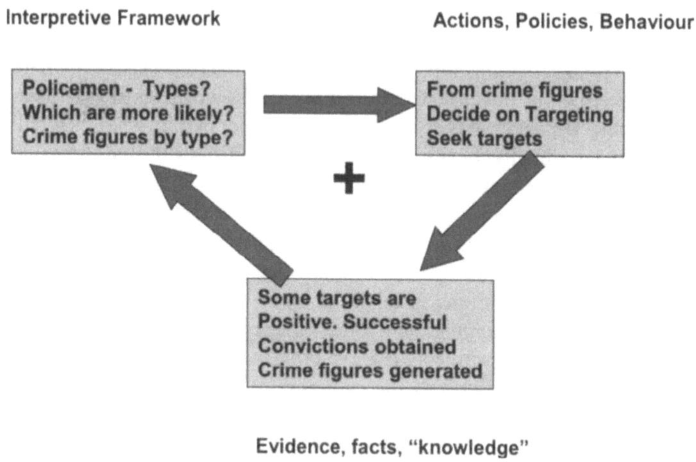

Fig. 4.6. Knowledge reflects experience and guides actions, which in turn generate knowledge. A positive feedback loop

In fact, policemen are not good scientists and should really test their theories all the time by performing random stop and search. If they did this, then they would find that their hypothesis was false. But, the performance pressures are there precisely to stop them performing "random" tests, and to make them get better results, not better knowledge. Even today, scientists are pressured into getting results rather than necessarily seeking the "truth," and so we can hardly blame policemen. The fact is that this loop of self-reinforcing learning is a very general property and in reality probably underlies much of what we call "tacit knowledge." Obviously, we learn most things from experience, and clearly science is not a necessary approach for many everyday tasks. But nevertheless, whenever we engage in attempting to "improve" our performance, we naturally attempt to identify which characteristics are significant indicators, and to use these as a guide to our actions. Experience is made up of an accumulation of such sequences, and many of them are probably useful. However, those made under pressure and formed and acted upon rapidly may well be nonsense.

The important point in our general understanding of social systems is that self-organising systems can break symmetry and create real behavioural differences between populations that were hitherto identical. This means that a system with internal diversity can evolve qualitatively over time, and this diversity doesn't even need to be real. It only needs to reside as a "speculation" in an actor's mind and that may be enough for non-linear interactions to amplify into a real presence.

Our example shows us that the evolutionary diversity of systems emerges from within them and can fashion their evolutionary pathway.

4.3 Manufacturing Evolution

The previous sections demonstrate theoretically how micro-diversity in character space, tentative trials of novel concepts and activities, will lead to emergent objects and systems. However, it is still true that we cannot predict what they will be. Mathematically we can always solve a given set of equations to find the values of the variables for an optimal performance. But we do not know which variables will be present, as we do not know what new "concept" may lead to a new structural attractor, and therefore we do not know which equations to solve or optimise. The changing patterns of practices and routines that are observed in the development of firms and organisations in an industry or business sector trace out an "evolutionary tree" of emergent new types of organisation. This is called a "cladistic diagram" (a diagram showing evolutionary history) and was originally developed in biology for describing the evolution of new species. Here however, we shall use it to show the history of successive new practices and innovative ideas in an economic sector. It generates an evolutionary history of both artifacts and the organisational forms that underlie their production (McKelvey 1982, 1994; McCarthy 1995; McCarthy, Leseure, Ridgway and Fieller 1997). In this case we can illustrate the ideas by considering manufacturing organisations in the automobile sector.

The organisational forms that have been identified are:

- Ancient craft system,
- Standardised craft system,
- Modern craft system,
- Neocraft system,
- Flexible manufacturing,
- Toyota production,
- Lean producers,
- Agile producers,
- Just in time,
- Intensive mass producers,
- European mass producers,
- Modern mass producers,
- Pseudo lean producers,
- Fordist mass producers,
- Large scale producers,
- Skilled large scale producers.

These are identified on the basis of their constituent "characteristics." In the case of the automobile sector the basic characteristics that have been observed are the different practices, routines and techniques that have emerged in the sector over time. The 16 different organisational forms listed above therefore differ from each other by being composed of different "bundles" of characteristics - defining therefore different "species" of organisation. This provides us with a clear definition of the "identity" of a particular company, in terms of constituent practices that make it up. Obviously, each of the 53 practices shown in Figure 4.7 has its own advantages that will add to the performance of the whole. However, in bringing in the ideas of "complexity" we are also concerned with the pair-wise interactions between each pair of practices, in order to examine the role of "internal coherence" on the organisational performance. In this "complex systems" approach, a new practice can only invade an organisation if it is not in conflict with the practices that already exist there. In other words, we are looking at "organisations" not in terms of simply additive features and practices, but as mutually interactive "complexes" of constituent factors.

Standardisation of Parts	1	TQM sourcing	27
Assembly Time Standards	2	100% inspection sampling	28
Assembly line layout	3	U-Shape layout	29
Reduction of Craft Skills	4	Preventive Maintenance	30
Automation (Machine paced shops)	5	Individual error correction	31
Pull Production System	6	Sequential dependency of workers	32
Reduction of Lot size	7	Line balancing	33
Pull procurement planning	8	Team Policy	34
Operator based machine maintenance	9	Toyota verification of assembly line	35
Quality circles	10	Groups vs. teams	36
Emloyee innovation prizes	11	Job enrichment	37
job rotation	12	Manufacturing cells	38
large volume production	13	Concurrent engineering	39
mass sub-contracting by sub-bidding	14	ABC Costing	40
exchange of workers with suppliers	15	Excess capacity	41
Training through socialisation	16	Flexible automation of product versions	42
Proactive training programmes	17	Agile automation for different products	43
Product range reduction	18	In-Sourceing	44
Automation (Machine paced shops)	19	Immigrant workforce	45
Multiple sub-contracting	20	Dedicated automation	46
Quality Systems	21	Division of Labour	47
Quality Philosophy	22	Employees are system tools	48
Open Book Policy with Suppliers	23	employees are system developers	49
Flexible Multifunctional workforce	24	product focus	50
set-up time reduction	25	Parallel processing	51
Kaizen change management	26	Dependence on written rules	52
		Further intensification of labour	53

Fig. 4.7. Fifty-three characteristics of manufacturing organisations

The analysis of the evolution of the different organisational forms is shown in Figure 4.8, which tells us which practices need to be added or taken away in order to derive the 16 different organisational forms.

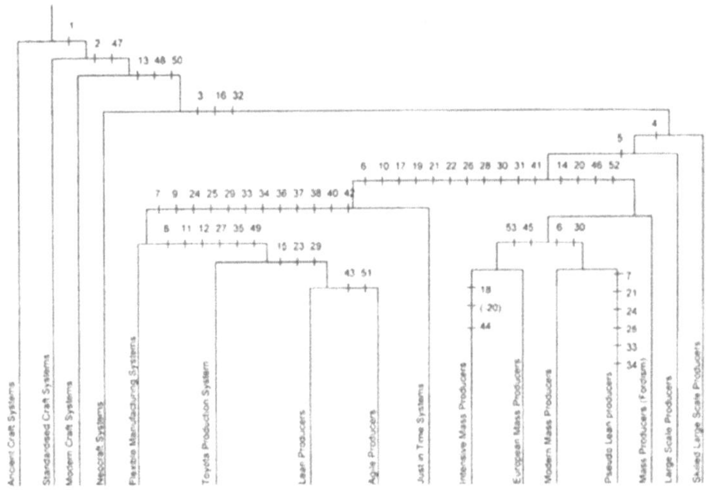

Fig. 4.8. The cladistic diagram for automobile manufacturing organisational forms (McCarthy, Leseure, Ridgway and Fieller 1997)

The idea behind the original work was that by understanding the "characteristic practices" that constitute the most effective organisations, it would be possible to provide precise advice on strategic changes that might allow a current "mass producer" to become a "lean and agile" one. However, it seems to us that the elimination of some practices and wholesale addition of many others is probably an impossible task, and that the only way to achieve it would be to replace an old factory working with the old rules with a brand new one working with the new ones. In fact the view we want to examine here is about the organisation as a complex system of interacting practices, and of successful organisational forms as reflecting the underlying synergy and coherence of its constituent practices. Using the ideas of complexity science, we want to understand this evolutionary tree and in so doing, understand the reasons behind the different organisational forms and their relative success. From a survey of manufacturers (Baldwin et al. 2003) concerning the positive or negative interactions between the different practices, a matrix of pair interaction was constructed allowing us to examine the "reasons" behind the emergent organisational forms, with successful forms arising from positive mutual interactions of constituent practices. This is shown in Figure 4.9.

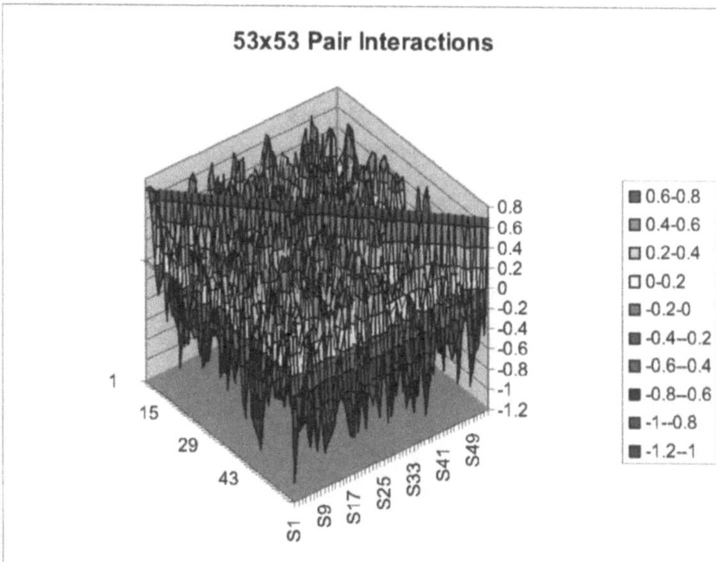

Fig. 4.9. The 53x53 matrix of pair interactions of the characteristic practices. It allows us to calculate the net attraction or conflict for any new practice depending on which ones are present already

We have then been able to develop an evolutionary simulation model, in which a manufacturing firm attempts to incorporate successive new practices at some characteristic rate. There is an incredible range of possible structures that can emerge, however, depending simply on the order in which they are tried. But, each time a new practice is adopted within an organisation, it changes the "invadability" or "receptivity" of the organisation for any new innovations in the future. This is a true illustration of the "path dependent evolution" that characterises organisational change. Successful evolution is about the "discovery" or "creation" of highly synergetic structures of interacting practices.

In Figure 4.10 we see the changing internal structure of a particular organisation as it attempts to incorporate new practices from those available. In the simulation, the number available start from the ancient craft practice on the left, and successively add the further 52 practices on the right. At each moment in time the organisation can choose from the practices available at that time, and its overall performance is a function of the synergy of the practices that are tried successfully. We see cases where practice 4, for example, is tried several times and simply cannot invade. However, practice 9 is tried early on and fails, but does successfully invade at a later date. The particular emergent attributes and capabilities of the organisation are a function of the particular combination of practices that constitute it.

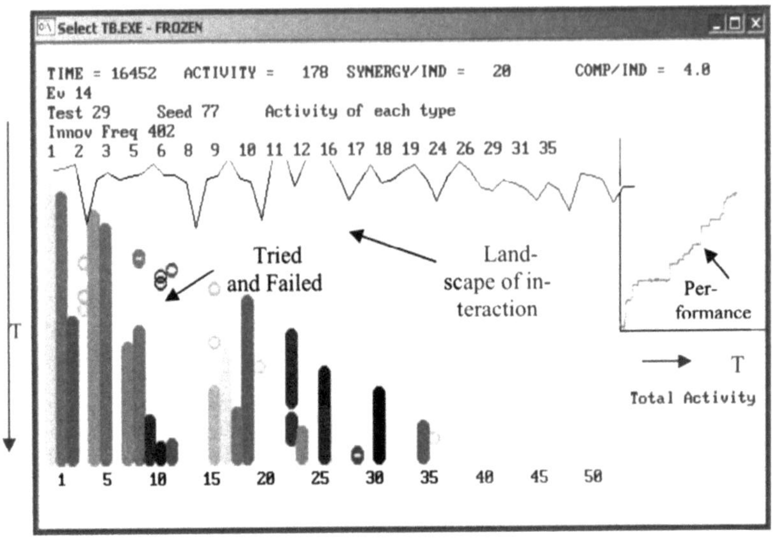

Fig. 4.10. An evolutionary model tries to "launch" possible innovative practices in a random order. If they invade, they change the "invadability" of the new system

The model starts off from a craft structure. New practices are chosen randomly from those available at the time and are launched as a small "experimental" value of 5. Sometimes the behaviour declines and disappears, and sometimes it grows and becomes part of the "formal" structure that then conditions which innovative behaviour can invade next.

Different simulations lead to different structures, and there are a very large number of possible "histories." This demonstrates a key idea in complex systems thinking. The explorations/innovations that are tried out at a given time cannot be logically or rationally deduced because their overall effects cannot be known ahead of time. Therefore, the impossibility of prediction gives the system "choice." In our simulation we mimic this by using a random number generator to actually choose what to try out, though in reality this would actually be promoted by someone who believes in this choice, and who will be proved right or wrong by experience, or in this case by our simulation. In real life there will be debate and discussion by different people in favour of one or another choice, and each would cite their own projections about the trade-offs and the overall effect of their choice. However, the actual success that a new practice meets with is pre-determined by the "fitness landscape" resulting from the practices already present and what the emergent attributes and capabilities encounter in the marketplace. But this landscape will be changed if a new practice does successfully invade the system. The new practice will bring with it its own set of pair interactions, modifying the selection criteria for further change. So, the pattern of what *could* then invade the system (if it were tried) has been changed by what *has already* invaded successfully. This is technically referred to as a "path dependent" process since the future evolutionary pathway is affected by that of the past.

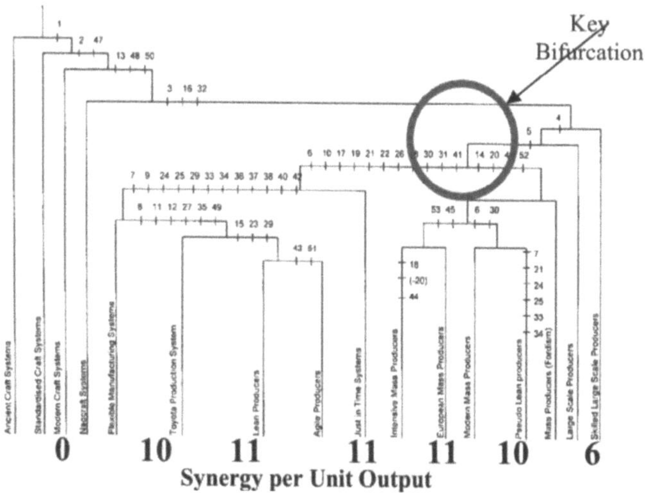

Fig. 4.11. The interaction matrix of Figure 4.9 allows us to show that the evolution from an ancient craft system, through mass production to lean and agile production is to organisations with increasing synergy

Our evolutionary model with output as in Figure 4.10 allows us to run through multiple evolutionary experiments of single firms, each one tracing a path through the possible futures by attempting to introduce new practices over time. The order in which they do this matters and so company performance and emergent capabilities reflects the luck and inspiration of their particular path. Because many firms are doing this simultaneously, the industry can be seen as a fuzzy cloud of experiments, some of which succeed, and some of which fail. This picture shows us that the evolution of a business sector is about the successive discovery of new features and practices and their successful incorporation into existing forms. Because outcomes are not known ahead of time, then individual companies adopt particular new practices according to their own subjective experience and knowledge, see Figure 4.12.

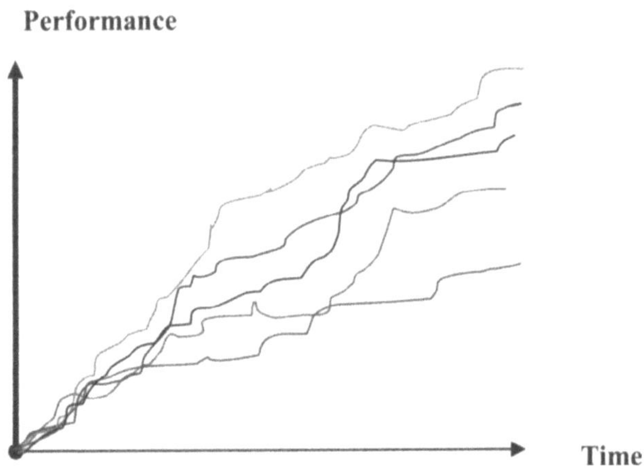

Fig. 4.12. Five different evolutionary experiments show that only some firms would survive. In an industry firms are constantly initiating innovations and some are failing

This leads to a divergent branching of possible forms as in Figure 4.11, and either to the successful adoption of a new idea or practice, or to its abandonment. A key bifurcation occurs between the "large scale producer" and the "just-in-time" system, as there are 17 conflicting practices that separate them. In other words, firms that take the "mass-producing" branches cannot really come back to the "lean and agile" ones after this. A tree of divergent forms is created from the survivors, and only new ideas that "fit" the internal organisation of a particular company - and in addition lead to successful performance in the market place - will be able to invade it. This allows us to make a much more subtle distinction concerning "best practices" since a practice that fits and improves one organisation may not fit or improve another. We now have a theory that allows us to understand and grasp the historical dependence of a company and its culture.

4.4 Structural Attractors

There are several important points about these results. The first is that the model above is very simple, and the results very generic. It shows us that for a system of co-evolving agents with underlying microdiversity and idiosyncracy, we automatically obtain the emergence of structural attractors such as the organisational forms shown in Figure 4.8. A structural attractor is the temporary emergence of a particular dynamical system of limited dimensions, from a much larger space of possible dynamical systems and dimensions. These are complex systems of interdependent behaviours whose attributes are on the whole synergetic. They have better performance than any single, pure homogeneous behaviour, but are less diverse than if all "possible" behaviours were present. In other words, they show

how an evolved entity will not have "all possible characteristics" but will have some that fit together synergetically, and allow it to succeed in the context that it inhabits. They correspond to the emergence of hypercycles in the work of Eigen and Schuster (1979) but recognise the importance of emergent collective attributes and dimensions. The structural attractor (or complex system) that emerges results from the particular history of search and accident that has occurred and is characteristic of the particular patterns positive and negative interactions of the components that comprise it. In other words, a structural attractor is the emergence of a set of interacting factors that have mutually supportive, complementary attributes.

What are the implications of these structural attractors?

Search carried out by the "exploratory" diffusion in character space leads to vastly increased performance of the final object. Instead of a homogeneous system, characterised by intense internal competition and low symbiosis, the development of the system leads to a much higher performance, and one that decreases internal competition and increases synergy.

The whole process leads to the evolution of a complex, a "community" of agents whose activities, whatever they are, have effects that feed back positively on themselves and the others present. It is an emergent "team" or "community" in which positive interactions are greater than the negative ones.

The diversity, dimensionality and attribute space occupied by the final complex is much greater than the initial homogeneous starting structure of a single characteristic practice or behaviour. However, it is much less than the diversity, dimensionality and attribute spaces that all possible behaviours or practices would have brought to the system. The structural attractor therefore represents a reduced set of activities from all those possible in principle. It reflects the "discovery" of a subset of agents/practices whose attributes and dimensions have properties that provide positive feedback. This is different from a classical dynamic attractor that refers to the long-term trajectory traced by the given set of variables. Here, our structural attractor concerns the emergence of variables, dimensions and attribute sets that not only coexist but actually are synergetic.

A successful and sustainable evolutionary system will clearly be one in which there is freedom and encouragement for the exploratory search process in behaviour space. Sustainability in other words results from the existence of a capacity to explore and change. This process leads to highly co-operative systems, where the competition per individual is low, but where loops of positive feedback and synergy are high. In other words, the free evolution of the different types of agent or practice, each seeking its own growth, leads to a system that is more co-operative than competitive. The vision of a modern, free market economy as being dominated by competition is false. Instead, competition leads to the emergence of co-operative structures and organisations with the capacity to continue to transform themselves over time.

The most important point really is the generality of the model presented above. If we think of an artefact, some product resulting from a design process, then there is also a parallel with the emergent structural attractor. A successful product or organisation is one in which the "bundling" of its different components creates emergent attributes and capabilities that assure the resources for its production and maintenance. However, the complication is that the emergent attributes and capabilities are not simply an additive effect of the components. If a change is made in the design of one component it will have multi-dimensional consequences for the emergent properties in different attribute spaces. Some may be made better and some worse. Our emergent structural attractor is therefore relevant to understanding what successful products and organisations are and how they are obtained. Clearly, a successful product is one that has attributes that are in synergy, and which lead to a high average performance. From all the possible designs and modifications we seek a structural attractor that has dimensions and attributes that work well together.

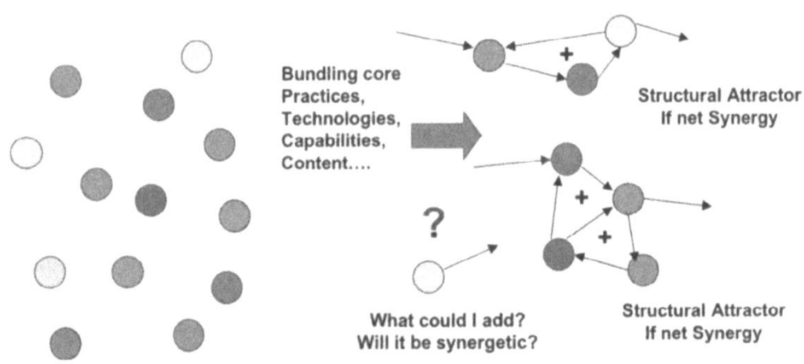

Fig. 4.13. On the left we have a "dictionary" of possible core concepts, practices or ideas. These are "bundled" on the right and if the different elements have synergy then the structure is successful

The structural evolution of complex systems, as shown in Figure 4.13, is about how explorations and perturbations lead to attempts to introduce modifications, and these lead sometimes to new "concepts" and structural attractors that have emergent properties. The history of any particular product sector can then be seen as an evolutionary tree, with new types emerging and old types disappearing. But in fact, the evolution of "products" is merely an aspect of the larger system of organisations and of consumer lifestyles that also follow a similar, linked pattern of multiple co-evolution.

4.5 Conclusion

The ideas explored above show how organisations such as firms explore possible functional innovations, and evolve capabilities that lead either to survival or to failure. They describe a divergent evolutionary diffusion into "possibility space." Each exploratory step is then either amplified or diminished depending on the "performance" of the products or services provided, which depends on the internal trade-offs within them, on the synergies and conflicts that it encounters or discovers in its supply networks, retail structures and in the lifestyles of final consumers.

Similarly, exploratory changes made in the supply network, in the retail structures, or in the different elements of the lifestyles of different types of individual all lead to a divergent exploration of possibilities. These are amplified or diminished as a result of the dual selection processes operating on one hand "inside them" in terms of the synergies and conflicts of their internal structures, and also "outside them," in the revelation of synergy or conflict with their surrounding features. So, a new practice can "invade" a system if it is synergetic with the existing structure, and this will then either lead to the reinforcement or the decline of that system in its environment if the modified system is synergetic or in conflict with its environment. Because of the difficulty of predicting both the emergent internal and external behaviours of a new action, the pay-off that will result from any given new action can therefore generally not be anticipated. It is this very ignorance that is a key factor in allowing exploration at all. Either the fear of the unknown will stop innovation, or divergent innovations will occur even though the actors concerned do not necessarily intend this. Attempting to imitate another player can lead to quite different outcomes either because the internal structure or the external context is found to be different.

Throughout the economy, and indeed the social, cultural system of interacting elements and structures we see a generic picture at multiple temporal and spatial scales in which uncertainty about the future allows actions that are exploratory and divergent, which are then either amplified or suppressed by the way that this modifies the interaction with their environment. Essentially, this fulfils the early vision of dissipative structures (Nicolis and Prigogine 1977), in that their existence and amplification depend on "learning" how to access energy and matter in their environment. Can they form a self-reinforcing loop of mutual advantage in which entities and actors in the environment wish to supply the resources required for the growth and maintenance of the system in question. In this way, structures emerge as multi-scalar entities of co-operative, self-reinforcing processes.

What we see is a theoretical framework that encompasses both the evolutionary and the resource-based theory of the firm. And, not only of the firm, but of the social and economic system as a whole. It is the complex systems dialogue between explorations of possible futures at one level, and the unpredictable effects of this both at the level below and the level above. There is a dialogue between the "trade-offs" or "non-linearities" affected inside and outside the particular level of exploration. But it is also true that all levels are exploring. Unless there is an imposition of rigid homogeneity up and down the levels of the system, there will necessarily be behavioural explorations due to internal diversity. And internal diversity can only be suppressed by an active selection strategy that immediately knows which entity will be effective, and which will not. But this is impossible, since the process is dynamic and takes time to register the relative performances. Because of this, diverse behaviours will invade the system and will coexist for considerable times, with selection operating only gradually. In this way, the multi-level systems are precisely the structures that can "shield" the lower levels from instantaneous selection, and allow an exploratory drift to occur, that can generate enough diversity to eventually discover a new behaviour that will grow. Without the multiple levels, selection would act instantly, and there would be no chance to build up significant deviations from the previous behaviour.

This paper sketches out an integrated theory of economic and social evolution. It suggests how the different types of people channel their needs into particular patterns of need for different products and services. These are delivered according to the non-linear interactions of synergy and conflict that lead to particular retailing structures, both expressing natural "markets" and within that complementarities between product categories and lines.

Products themselves exist as embodiments of attributes that cluster synergetically and different product markets emerge naturally as a result of inherent conflicts between attributes. For example, a palmtop computer cannot have a really easy to use keyboard (under existing design concepts) and so notebooks and laptops exist in a different market to palmtops. Similarly, toasters and telephones also occupy separate markets because answering a call on a toaster/telephone can set your hair on fire. So, again it is the "complementarities and conflicts" of possible attributes that structures the space of possible product or service markets.

On the supply side, the capabilities of organisations, and the products and services that they create, are the result of a creative evolutionary process in which clusters of compatible practices and structures are built up, in the context of the others, and discover and occupy different niches. At each moment, it is difficult to know the consequences of adopting some new practice, since the actual effect will depend on both the internal nature of the organisation and its actual context and relationships it had developed. For this reason there is no such thing as "best practice," and introducing new ideas is bound to be an exploratory, risky business. In the short term it will always be better to simply optimise what already exists, and not to risk engaging on some innovation. But over time, if a firm does not engage in evolution then extinction becomes not only possible but certain.

The whole system is an (imperfect) evolutionary, learning system in which people learn of different ways that they could spend their time and income, and what this may mean to them. Companies attempt to understand what customers are seeking, and how they can adapt their products and services to capture these needs. They attempt to find new capabilities and practices to achieve this, and create new products and services as a result. These call on new technologies and materials and cause evolution in the supply networks. Technological innovation, cultural evolution and social pressures all change the opportunities and possibilities that can exist, and also the desires and dreams of consumers and their patterns of choice and of consumption.

We live in a world of "restless capitalism" (Metcalfe 1998, 1999) that is capable of exploring and creating new needs and behaviours, and in doing so means that we live with continuing and necessary uncertainty that is rooted in the essential complexity and non-transparency of the world. This imperfect learning process means that decisions will tend to reflect the short-term positive performance of something with respect to the dimensions of which we are aware, but obviously, in a complex system there will be all kinds of less obvious factors that are perhaps adversely affected, perhaps over the longer term, but even quite immediately. In other words, what we choose to do is dependent on "what we are aware of," and so system changes reflect our limited understanding of what will actually affect us. This is because our actions are based on our limited understanding and knowledge of the complex systems we inhabit. And their evolution therefore bears the imprints of our particular patterns of ignorance. So, we may grab economic gain, by pushing "costs" into the "externalities," or we may seek rapid satisfaction from consuming some product that actually harms us, or our community, or our region, or the ozone etc. over the longer term.

Complex systems thinking is telling us that we are forever at risk of evolving into an unknown future, with sometimes interesting, sometimes painful consequences. Though we construct edifices of routine and regularity into our system, the need to innovate and change will always assert itself at some point in time, and we shall be forced to move to a new, temporary set of routines and regularities. Rational thinking has revealed the limits to rational thinking and we see that evolution springs exploration, which is allowed by uncertainty, and uncertainty results from creative evolution. Uncertainty and surprise are a necessary feature of life itself.

This work was supported by the ESRC NEXSUS Priority Network.

References

Allen PM (1976) Evolution, population dynamics and stability. Proc Nat Acad Sci, USA, Vol 73, No 3, pp 665-668

Allen PM (2000) Knowledge, ignorance and learning. Emergence 2(4):78-103

Allen PM (2001) Knowledge, ignorance and the evolution of complex systems. In: Frontiers of evolutionary economics: Competition, self-organisation and innovation policy. Cheltenham, UK

Allen PM (2001) A complex systems approach to learning adaptive networks. International Journal of Innovation Management 5(2):149-180

Baldwin JS, Allen PM, Winder B, Ridgway K (2003) Simulating the cladistic evolution of manufacturing. Innovation: Management, Policy and Practice 5(2,3):144-156

Eigen M, Schuster P (1979) The hypercycle. Springer, Berlin

McCarthy I (1995) Manufacturing classifications: Lessons from organisational systematics and biological taxonomy. Journal of Manufacturing and Technology Management - Integrated Manufacturing Systems 6(6):37-49

McCarthy I, Leseure M, Ridgeway K, Fieller N (1997) Building a manufacturing cladogram. International Journal of Technology Management 13(3):2269-2296

McKelvey B (1982) Organizational systematics. University of California Press, California

McKelvey B (1994) Evolution and organizational science in evolutionary dynamics of organizations. Oxford University Press, Oxford

Metcalfe JS (1998) Evolutionary economics and creative destruction. Routledge, London

Metcalfe JS (1999) Restless capitalism, returns and growth in enterprise economics. CRIC, University of Manchester

Nicolis G, Prigogine I (1977) Self-organization in nonequilibrium systems. Wiley-Interscience, New York

5 Managing the Unexpected: Complexity as Distributed Sensemaking[1]

Karl E. Weick[2]

In 1998 the Centers for Disease Control (CDC) published a statement of their strategy entitled "Preventing Emerging Infectious Diseases: A Strategy for the 21st Century." They described their central challenge this way: "because we do not know what new diseases will arise, we must always be prepared for the unexpected" (p. vii). Soon after they published that statement CDC was confronted with an unexpected emerging disease, the West Nile Virus, which they misdiagnosed initially.

Much of what we think of as crucial in organizational life is visible in this incident. The question is, to what extent do concepts dealing with complexity help us understand what is visible in this incident? The juxtaposition of the concept of complexity and the activity of diagnosing sets up a tension that was anticipated by Immanuel Kant when he said, 'perception without conception is blind, and conception without perception is empty.' Do concepts associated with complexity remove blindness when we watch how CDC wades into a puzzling set of symptoms? And do observations of diagnostic activity that unfolded at CDC remove some of the emptiness associated with ideas of non-linear dynamics, emergence, turbulence, complex adaptive systems, heterogeneous agents, self-organization, and messes?

I do not intend to interweave complexity theory with organizational theory as is already being done by people like Anderson (1999) and Eisenhardt and Bhatia (2002). Instead, I want to talk about organizing in the face of the unexpected. I want to use the West Nile episode as my running illustration and I want to juxtapose ideas about complexity, cognition, and sensemaking in order to argue that if complexity ideas[3] are made more cognitive and more relational, they look like

---

[1] Investigation of the West Nile incident is an ongoing joint project that involves Joe Porac, Huggy Rao, and Karl Weick, with the assistance of Katherine Lawrence. I am indebted to my collaborators for ongoing discussions that have helped all of us see the larger significance of this incident for organizational theory.

[2] Rensis Likert College Professor of Organizational Behavior and Psychology, Professor of Psychology, University of Michigan, karlw@bus.umich.edu.

[3] My synoptic view of complexity theory would include the following. Complexity 'theory' essentially is a collection of intuitions about complexity grounded in computer demonstrations of bit strings interacting, chemical clocks, fluid dynamics, weather, and lightning. The central derivative social insight that tends to be used in organization theory is this: "partially connected agents operating within simple rules drive complicated adaptive behavior at the system level" (Eisenhardt and Bhatia 2002:462). The essential shift in or-

human sensemaking. And if you make that translation, then complexity ideas would have even more relevance to human organizing.

5.1 Overview of the Event

The basic story of the West Nile diagnosis is this.[4] The Centers for Disease Control (CDC) were contacted on August 27, 1999 by the NYC Health Dept., and formally invited on August 30, 1999, to help diagnose a cluster of patients who had been admitted to intensive care at Flushing Hospital with unusual symptoms: fever, headache, mental confusion, severe muscle weakness. Some of these admissions died. Among the suspected causes were botulism (a potential bioterrorism agent), and Guillain Barre disease. But analysis of spinal fluid had also suggested a viral infection. After testing samples of blood serum, CDC and NYC jointly announced on September 3 that there was an outbreak of mosquito borne St. Louis Encephalitis (SLE). An intense mosquito eradication program was initiated by the Giuliani administration within 2 hours.

The initial picture of SLE had several loose ends, however. For example, New York State Laboratories also analyzed serum samples using two different tests. With serological tests they found evidence supporting a diagnosis of SLE. But with PCR test they found evidence inconsistent with a diagnosis of SLE (GAO 2000:44). CDC's Vector Borne laboratory in Fort Collins, Colorado used a third type of test, Elisa. An Elisa test is a blunt instrument in the sense that it identified the family of the suspect virus (a flavivirus) but not the specific virus itself (Gill 2000:9-11). Since almost all of the 70 varieties of virus in this family were alien to North America, this seemed to pose no problem. CDC announced that they found "a reaction characteristic of SLE" and that SLE was the "most likely cause." Lost in this diagnosis was evidence contrary to the diagnosis. SLE is not associated with muscle weakness, or with local outbreaks only, nor does it affect birds and horses.

At the same time that humans were dying, an increasing number of birds were dying in the NYC area. A staff person in the NYC health department phoned CDC on September 4 suggesting that there might be a bird-human connection (GAO, 2000:46). But since SLE, the announced diagnosis, does not kill birds, CDC saw these bird deaths as merely coincidental. People concerned with wildlife, domestic animals, and zoo animals were less certain than CDC that the deaths were unconnected. Repeated testing within the animal community began to confirm that birds

ganization theory is that complexity is *not* viewed as a structural variable. Complexity analyses focus on systems pushed away from equilibrium.

[4] Although we are currently interviewing people connected with the West Nile diagnosis, all references to the West Nile in this Chapter come from readily available public documents. Key sources that I used include Steinhauer and Miller 1999, Gill 2000, Government Accounting Office 2000, Scott 2002, Hall 2003, Asnis et al. 2000, Wadler 1999, Steinhauer 1999, Despommier 2001, White and Morse 2000, and Drexler 2002.

were dying from a virus other than SLE, and it was a virus that no one could identify. For example, birds had been dying at the Bronx zoo, a facility located close to the area where the majority of the human victims lived. By August 25 the bird deaths had become a concern to pathologist Tracey McNamara at the zoo. And by September 9, she had contacted CDC for help. CDC did not return her call, so McNamara began to activate her own network of animal laboratories to examine the samples of zoo deaths. She was worried about a danger that directly straddled the human-animal connection, namely, one of her technicians who was doing necropsies on the birds had suffered a needle stick injury. That could have serious health consequences. Thus, CDC knew the possibility of a bird-human connection almost from the beginning. But if such a connection were taken seriously, then this meant that their initial public diagnosis was wrong, since SLE does not kill birds.

In truth, their initial diagnosis was wrong. On September 23, three weeks after announcing that NYC was experiencing an outbreak of SLE, it was re-announced that NYC was actually experiencing an outbreak of a virus never before seen in the New World, a virus called West Nile. Other laboratories at Fort Dietrich, Ames Iowa, and UC Irvine converged on this finding shortly before CDC did.

From an organizational standpoint, what is interesting about this incident is that even though CDC tried to expect the unexpected, they wound up expecting the expected. Faced with an emerging disease, CDC initially saw a well-established disease. That slip-up had ominous overtones for many who viewed this episode as a dress rehearsal for how well the U.S. could cope with bioterrorism. The post mortems on the event were predictably varied and included statements such as, "CDC officials didn't do anything wrong, but they did not do all the right things" (Gill 2000:22); "CDC had tunnel vision and should have had a more open-minded approach" (Gill 2000:22).

What we have here is an organization, CDC, with a reputation for reliable accuracy that gets it wrong, with eight million New Yorkers looking on, while other local, state, and federal organizations have different hunches of the right answer. A closer look at this incident affords a chance to explore what it means to work at the edge of codified knowledge, using distributed cognition, in an effort to make sense and save lives. Here's where complexity comes in. I want to start with the working assumption that "The cognitive properties of human groups may depend on the social organization of individual cognitive capabilities" (Hutchins 1995:176). Thus, if we spot flaws in collective induction, then we may find an explanation for their genesis in the way people are organizing. Stated more compactly, the degree of intelligence manifest by a network of nodes may be determined by the quality, not just the quantity of its interconnectivity (Taylor and Van Every 2000:213).

5.2 Organizations are Loosely Connected

Like many people writing about complexity, I start with the assumption that organizing emerges among agents who are loosely connected. A loosely connected organization looks something like the picture that Pfeffer and Salancik drew:

> An alternative perspective [to that of the rational organization] on organizations holds that information is limited and serves largely to justify decisions or positions already taken; goals, preferences and effectiveness criteria are problematic and conflicting; organizations are loosely linked to their social environments; the rationality of various designs and decisions is inferred after the fact to make sense out of things that have already happened; organizations are coalitions of various interests; organization designs are frequently unplanned and are basically responses to contests among interests for control over the organization; and organization designs are in part ceremonial. This alternative perspective attempts explicitly to recognize the social nature of organizations. (1977:18-19)

In order to better adapt that image to complexity thinking, we can describe organizations as social order where "Groups[5] composed of individuals with distributed-segmented, partial-images of a complex environment can, through interaction, synthetically construct a representation of it that works; one which, in its interactive complexity, outstrips the capacity of any single individual in the network to represent and discriminate events....Out of the interconnections, there emerges a representation of the world that none of those involved individually possessed or could possess" (Taylor and Van Every 2000:207). The basic theme implied by this statement is that variations in interconnection produce variations in the representations that are synthetically constructed. This suggests again that different forms of network have different cognitive consequences. Some network forms may produce ignorance, tunnel vision, and normalizing, whereas other forms may produce novel insights, original syntheses, and unexpected diagnoses.

5.2.1 Loosely connected systems can be variously organized.

In order to conceptualize network forms in a way that juxtaposes cognition, complexity, and organizing, we can talk about distributed problem solving using the classic ideas proposed by James Thompson (1967). He suggests that work, such as distributed information processing, tends to exhibit three forms of task interdependence that lend themselves to three forms of coordination. Our proposal is that these forms of task interdependence also induce distinct forms of cognitive interdependence. Thompson distinguishes among pooled interdependence that is coordinated by standardization, to which we add the possibility that this form induces skill-based action[6] and automatic cognition[7]; sequential interdependence coordi-

[5] We could substitute the word 'networks' for the word 'groups' and this image still works.

[6] The three levels of action were described by Rasmussen (1983). Skill-based behavior represents sensori-motor performance during acts or activities that, after a statement of an intention, take place without conscious control as smooth, automated, and highly inte-

nated by plan, to which we add the possibility that this form induces rule-based action and heuristic cognition built around recipes (memorized rules, if-then); and reciprocal interdependence coordinated by mutual adjustment to which we add the possibility that this form induces knowledge-based action and controlled cognition. All three forms can co-exist, and Thompson treated the three as if they were a Guttman scale. Reciprocal interdependence presumes the existence of pooled and sequential.

If you have an emerging, unexpected infectious disease, it is most likely to be detected by controlled cognition. But, in the West Nile episode, in the early stages, there appears to be coordination by standardization and pooled task interdependence. The task of analyzing samples is partialed out among laboratories, the laboratories run their tests, and they send the results to CDC. "Each part renders a discrete contribution to the whole and each is supported by the whole" (Thompson 1967:54).

The piece I want to add is that the organization of the workflow can affect the way people think. The cognitive interdependence in the early stages of West Nile looks like pooled workflow interdependence in the sense that different people have different pieces of information and they contribute those pieces for assembly into a meaningful diagnosis. The problem is, mere assembly does not guarantee meaning. Each part is meaningless until it is related to some other part whose meaning, in turn, is dependent on the meaning of the initial part. Making meaning is an iterative process. Recall that what we are dealing with in the West Nile event

---

grated patterns of behavior. In rule-based behavior a sequence of subroutines in a familiar work situation is typically consciously controlled by empirical cue-action correlation. The person is aware that alternative actions are possible and has to make a choice. During unfamiliar situations for which no know-how or rules for control are available, the control moves to a higher conceptual level, in which performance is goal-controlled and knowledge-based. Viewed as a hierarchical control structure, the skill-based level represents the continuous real-time control of activities, the rule-based level reflects the adaptive choice among preplanned decision rules and the knowledge-based level reflects intelligent self-organization of behavior.

[7] George Mandler (1984) describes forms of cognition this way: "As a first approximation, I assume that actions and thoughts that issue from automatic processes require no intention or choice. Nonautomatic actions and thoughts are 'conscious' - they may have equivocal outcomes, they sometimes require intentions, and, in particular, they usually require choices, decisions, and selections....The well-known phenomenon that skills are conscious when first acquired but become unconscious once they are well practiced describes the availability of choices in the form and the inevitability of outcomes in the latter kind of performance. It should be obvious that selective-search mechanisms require access to schemas that are activated. Activation is thus a necessary condition for selective retrieval, but it is both necessary and sufficient for the occurrence of automatic thoughts and action....Another way of distinguishing automatic and nonautomatic processes is to refer to them as nondeliberate versus deliberate. Deliberate retrieval implies that the target or goal state is not immediately available, that choice need to be made and searches and selection initiated." For a more detailed discussion of the automatic-controlled duality, see Chaiken and Trope (1999).

is an emerging disease, a non-routine problem, equivocal cues, and ambiguity. Pooled task interdependence won't generate the reciprocal *cognitive* interdependence that is needed to reduce the ambiguity of the strange cluster of symptoms. Pooled interdependence is the interdependence of routines and standardization in work; but pooled workflow interdependence is also the cognitive interdependence of stereotypes, confirmation, codification, and automatic thinking. That is precisely the form of cognition that is *not* suited to detect emerging diseases.

To see this more clearly, think about the tendency of people to normalize the unexpected, as happened for example in the events leading up to the Challenger disaster. People often handle the unexpected by ***normalizing*** it out of existence.[8] The temptation to do this should be especially strong when the disease is "emerging" since, taken literally, something that emerges resembles its neighbor quite closely in its early stages. As it emerges more fully and becomes more distinct, it is less likely to be confused with its neighbor. Notice also that, in the beginning stages, you don't know that it is an emerging disease. It looks more like a variant of an old disease, and this is an ideal situation for fixation of attention and a failure to revise a situation assessment as new information comes in (see Cook and Woods 1994:274-277 on 'fixation problems'). Therefore, if you want to prepare for the unexpected, then you have to weaken or neutralize the tendency to normalize. You have to encourage ambivalence. You have to question your associates and argue with them, even though the paradigm is underdeveloped (remember, people are working at the edge of codified knowledge). You have to think in a more mindful, less automatic manner. You have to engage in controlled thinking that is more commonly associated with doubt, inquiry, argumentation, and deliberation. That is the thinking of reciprocal interdependence and coordination by mutual adjustment.

There were moments of reciprocal interdependence among animal laboratories in the West Nile incident, and these seemed to hasten the realization that people in NYC were dealing with an anomalous virus. Moments of reciprocal interdependence and controlled cognition were less frequent on the human side where "inquiry" basically took the form of routine diagnosis to see if sick people had the known SLE virus. Less common was the question, does the initial diagnosis remain viable and what symptoms remain inconsistent with it?

The basic point is that forms of task interdependence may induce forms of cognitive interdependence that hinder solution of the presenting problem. For example, if the problem is non-routine and requires controlled thinking, and if the task interdependence is pooled, then the task may induce automatic, skill-based thinking which is better suited to routine problems. If the non-routine problem is treated as if it were routine, then a puzzling member of the flavivirus family may well be interpreted to be familiar member of the family, namely, SLE.

The tricky part of a multi-organization network is that any one group may be capable of all three forms of task interdependence and all three forms of cognitive interdependence. When groups are strung together in a network, however, the network itself tends to be dominated by a single form of interdependence, either

[8] See William James (1992:512 ff) for an insightful description.

pooled with a central assembler, sequential with progressive assembly, or reciprocal with joint assembly. The problem with network structures is that reciprocal interdependence is most readily achieved on a local basis among small sets of players. As more subsets are hooked together, the interdependence drifts from reciprocal to sequential to pooled. Coincident with this drift is a shift from controlled cognition to heuristic cognition and finally to automatic cognition. If the network is faced with a non-routine problem, and if controlled collective cognition is weakened and replaced by collective cognition that is more automatic, then network failure is more likely. Networks may be faulty forms for emerging problems unless they are managed mindfully. This line of analysis predicts that a disproportionately large number of network failures occur when problems require controlled thinking (i.e., the presenting problem is ambiguous, equivocal, confusing). Failures occur because the pooled and sequential interdependence that is typical of networks induces inappropriate modes of thinking. Automatic thinking is imposed on problems that require controlled thinking.

5.3 Collective Cognition Affects Sensemaking

I now want to enlarge the analysis and bring in the theme of sensemaking which "involves turning circumstances into a situation that is comprehended explicitly in words and that serves as a springboard into action" (Taylor and Van Every 2000:40). Sensemaking is a diagnostic process directed at constructing plausible interpretations of ambiguous cues that are sufficient to sustain action. Interorganizing, thus, is understood as a cue interpretation process that requires cognitive coordination in the interest of wise action.

While self-organizing and emergence and co-evolution are crucial concepts for complexity theorists, when it comes to organization it is crucial that we also not lose sight of the *reactive* quality of organizations. This property is clearly visible in the West Nile episode and in the way CDC operates. There is often no good way to anticipate the next disease outbreak short of waiting for a few people to get sick. Henig (1993) asks, "What is the next AIDS?" Her answer, "You can't do much until the first wave of human infection occurs. You can't prevent the next epidemic. Furthermore, signs get buried among other diseases. If you find a new virus, you don't know whether it is significant or not until a human episode occurs. The trouble is that by the time you do establish that it is significant, the virus has already settled into hosts, reservoirs, and vectors and is being amplified. Edwin Kilbourne, a microbiologist at Mt. Sinai hospital states the reactive quality of diagnosis: "I think in a sense we have to be prepared to do what the Centers for Disease Control does so very well, and that is put out fire…It's not intellectually very satisfying to wait to react to a situation, but I think there's only so much preliminary planning you can do. I think the preliminary planning has to focus on what you do when the emergency happens: Is your fire company well drilled? Are they ready to act, or are they sitting around the station house for months" (Henig

1993:193-194). Notice that in a reactive world, a highly refined planning system is less crucial than the capability to *make sense out of an emerging pattern.*

There are several sensemaking puzzles in the West Nile incident including: Is this bioterrorism?, Is this botulism?, I've never seen muscle weakness associated with brain inflammation before, SLE shouldn't be in NYC, these profiles of SLE actually look "borderline," why are flamingos dying but emus in the next cage thriving?, I have never seen brain lesions that are this severe.

The dynamics of sensemaking[9] have some subtle properties. These subtleties were described by the late Paul Gleason, one of the best wildland firefighting commanders in the world. Gleason felt he was most effective as a leader when he viewed his job as one of sensemaking rather than decision making. In his words, "If I make a decision it is a possession, I take pride in it, I tend to defend it and not listen to those who question it. If I make sense, then this is more dynamic and I listen and I can change it. A decision is something you polish. Sensemaking is a direction for the next period."

When Gleason perceives himself as making a decision, he reports that he postpones action so he can get the decision "right" and that after he makes the decision, he finds himself defending it rather than revising it to suit changing circumstances. Both polishing and defending eat up valuable time and encourage blind spots. If, instead, Gleason perceives himself as making sense of an unfolding fire, then he gives his crew a direction for some indefinite period, a direction which by definition is dynamic, open to revision at any time, self-correcting, responsive, and with more of its rationale being transparent.

5.3.1 Complexity and cognition as sensemaking

Earlier we described the organizing of West Nile Virus as socially distributed cognition among interdependent players with differing priorities and local resources. Socially distributed cognition can be analyzed as a structural problem of task interdependence and coordination as we just saw. But it can also be analyzed as a set of socially organized resources for sensemaking. Here we focus on a different set

[9] The challenge for people at CDC is a lot like the challenge that faces incident commanders at the scene of a disaster. As Rhona Flin says, Incident commanders face

1. extremely difficult decisions
2. ambiguous and conflicting information
3. shifting goals
4. time pressure
5. dynamic conditions
6. complex operational team structures
7. poor communication
8. every course of action carries significant risk (1996:37).

Their challenge is to continually make sense of an unexpected and dynamic situation that is characterized by unfamiliarity and scale and speed of escalation (paraphrase from Flin 1996:105).

of issues. Now we ask whether *social resources at CDC were organized to create a plausible story that was actively updated through ongoing attention to shifting patterns of cues*. This shift from a structural analysis to a more processual analysis aligns us even more closely with complexity ideas. Seven different resources for sensemaking are implied by this description, and they are captured by the acronym SIR COPE.[10]

Social. "S" stand for social, and captures the fact that organizational sensemaking is interactive, relational, and in Eric Eisenberg's (1990) words, consists of "coordination of action over alignment of cognitions, mutual respect over agreement, trust over empathy, diversity over homogeneity, loose over tight coupling, and strategic communication over unrestricted candor" (p. 27).

The crucial idea here is that intelligence is a product of interconnectivity (Taylor and Van Every 2000:213). Interconnectivity and its role in cognition and sensemaking can be depicted more formally in terms of the concept of heedful interrelating. The basic idea in heedful interrelating is that a collective mind capable of varying degrees of intelligence emerges as a kind of capacity in an ongoing activity stream when activities among people are tied together as contributions that constitute and are subordinated to a joint system (Weick and Roberts 1993). The mind is more fully developed if those interrelations occur with greater heedfulness.[11]

---

[10] The seven resources, captured by the acronym SIR COPE, vary along dimensions whose anchor points can be labeled thusly:

1. social-solitary resources
2. defined-vague identity
3. backward-forward noticing
4. equivocal-confirmed cues
5. continuous-episodic flow of events
6. possibility-probability as criterion for narratives
7. enactive-reactive as form of action.

Sensemaking resources characterized by the left-hand terms are presumed to be more effective in reducing ambiguity than are sensemaking resources characterized by the right hand terms. This line of argument predicts that in the early stages of the West Nile episode, the sensemaking resources that CDC directed at the problem were toward the right hand end of each of these dimensions. As the pattern of resources began to move more toward the left-hand end of each dimension, people were better able both to notice anomalies that didn't fit and to invent a newer story into which they did fit.

[11] An example of less heedful interrelating is Winston Churchill's reconstruction of why he failed to see that Singapore was vulnerable to land invasion in World War II. Allinson (1993) notes that "A good illustration of the awareness of multiple causality may be found in Churchill's response to his horrified discovery that Singapore, rather than being impregnable, proved to be highly vulnerable to a Japanese land invasion. In his history of World War II, Churchill comments: "I ought to have known. My advisors ought to have known and I ought to have been told, and I ought to have asked" (p. 11). What is crucial here is that all four lapses are lapses of interconnection. Participants are not attentive to their contributions, representations, and subordination to a possible emerging system for gathering information about unexpected events.

Identity. Sensemaking unfolds from some standpoint, some frame of reference, some identity. Several potential identities are at work in the West Nile incident. These include CDC as "detective," "expert," "public health guardian," a "reference lab" for World Health Organization (WHO), the go-to unit when diagnosis gets tough (akin to the wildland firefighting crews of hotshots), and the expert at shoe-leather epidemiology (Last 2001:168). CDC's identity is less that of an "integrator" where the network becomes the expert. Furthermore, CDC's claimed identity as a site that practices "basic science" makes the issue of misidentification less clearcut. Stephen Ostroff, a central player in the West Nile incident, was quoted in *The New York Times* as saying, "This [West Nile Virus] was not a mistake. This is how science proceeds in outbreak investigations. Confusion is a normal part of an emerging disease investigation" (Steinhauer and Miller 1999).

Retrospect. Action is always just a tiny bit ahead of cognition. We always see a little too late what we have done and what its consequences are. For example, the Annual Report for 1999 published by Applied Energy Systems (AES) contains this statement: "Strategy is typically developed through a series of business experiments carried out by our people as they seek to achieve that purpose [serve the electricity needs of the world]. Describing strategy, then, is more of a retrospective look at what has happened than a road map to the future" (p. 39). Applied to issues of diagnosis, retrospective thinking is understood as belated understanding of what one illness or condition one was facing back then, though didn't realize it at the time. Marianne Paget (1988) is quite insightful on this point: "A mistake is situated in the conduct of medical work. It is discovered in the aftermath of action and activity, in reflection about medical action. 'I made a mistake. If I knew then what I know now I would have done x, but I did not know then. If I had it to do all over again I would do x, but I do not always have it to do all over again. I mistook x for y. Was I distracted? Was I 'misled' by the patient?" (p. 124). Physicians don't count errors that occur in diagnosis and therapy as errors. Instead, "they count them as progressive approximations of their understanding of the character of illness" (p. 137). This may be one reason interviewers get blank stares when they say, "let's talk about medical errors." The notion of approximations and updating is a crucial aspect of retrospect. "The work process unfolds as a series of approximations and attempts to discover an appropriate response. And because it unfolds this way, as an error-ridden activity, it requires continuous attention to the patient's condition and to reparation" (p. 143). In other words, the risk of medical action, including outbreak diagnosis, is often exposed retrospectively.

Cues. Part of the problem in the West Nile incident is that CDC is comparing cues in 1999 with outdated information about West Nile Virus. The misidentification occurs because of a close family resemblance between new inputs and older indicators. When you work at the edge of codified knowledge, with an outdated classification system, then you work with vague equivocal cues. And you may or may not know that this is the case. The newer cues are, in Diane Vaughan's (1996) image, weak, mixed, routine.

Ongoing. Crows are dying, people are dying, people are calling with questions and crow sightings, samples are piling up on the loading dock, some of them better labeled than others. Malathion is being sprayed, an election campaign for the senate is being waged between Giuliani and Hilary Clinton, and there are suspicions of bioterrorism, all in the context of emerging infectious diseases. Any interruption in an ongoing project creates either a prompt repair and recovery, or a detached, atomistic analysis. The goal is to stay in the action because, once you pull away and adopt a detached atomistic view, you lose context, information, situated cognition, and tools made meaningful by actual use.

Plausibility. The initial story says that there is a high probability that NYC is faced with an outbreak of SLE that is spreading. This story is incomplete, is based on selected data, but it also triggers action and potential new inputs that could revise the initial story. Plausibility gets people in action, which is helpful when accuracy is a moving target. The environment continues to change, and action based on the SLE diagnosis stirs up new puzzles. Fort Collins begins to see that their positive readings for a SLE reaction are weak ("borderline"), and that there is a stronger reaction for West Nile Virus. A fuller story needs to be crafted. If people fixate on their first plausible story and stop there, then they do have a sense of sorts, but one that holds together only if newer cues and consequences are ignored.

Enactment. Nigel Nicholson (1995:155) has described enactment in the following way: enactment is a concept developed "to connote an organism's adjustment to its environment by directly acting upon the environment to change it. Enactment thus has the capacity to create ecological change to which the organism may have subsequently to adjust....Enactment is thus often a species of self-fulfilling prophecy....One can expect enactment processes to be most visible in large and powerful organizations which have market-making capacity, but they are no less relevant to the way smaller enterprises conceive their contexts and make choices about how they will act in relation to them." Examples of enactment include physician-induced disease (iatrogenic) which when diagnostic tests or lines of questioning create sickness that was not present when the patient first consulted with a physician; an air traffic controller who creates a *holding pattern* by stacking several aircraft in a small area of airspace near a busy airport and, in doing so, enacts a cluttered display on the radarscope that is more difficult to monitor; or organizations that encourage *closeness to the client* enact a permissive world that encourages outrageous customer demands that can only be remedied by firing the client they tried so hard to recruit. In each case, individual work can enact conditions that other people and other systems have to cope with. For example, iatrogenic disease does not stop at the physician's door as the newly troubled patient walks out. Instead, the altered patient walks into the medical care system where the consequences of the initial treatment spread and where the patient's problems with the physician become other people's problems as well. Enactment creates contingencies as well as events. The initiating conditions seem small in comparison to macro events only because these examples articulate the local turning point, the point of bifurcation, the moment of initiation. These triggering moments often serve to implant small but uncontained outcomes in larger systems. These embedded, uncontained outcomes continue to grow undetected until they spawn unanticipated consequences that threaten legitimacy, competence, and control.

5.4 Conclusions

I have argued that how you are organized (the 'social' dimension in sensemaking) determines the depth of your resources for sensemaking. Organizing is about workflow interdependence, cognitive interdependence, and the intelligence enacted by the way the interconnection occurs. The quality of that interconnectivity, the degree of heed involved in interrelating, affects the quality of representations constructed by members. Recall the earlier image of organizations as "Groups composed of individuals with distributed - segmented, partial-images of a complex environment can, through interaction, synthetically construct a representation of it that works; one which, in its interactive complexity, outstrips the capacity of any single individual in the network to represent and discriminate events....Out of the interconnections, there emerges a representation of the world that none of those involved individually possessed or could possess" (Taylor and Van Every 2000:207). Different "interaction" produces different "synthetic construction of a representation." When interrelating is less heedful,[12] there tends to be more normalizing, more susceptibility to the fallacy of centrality,[13] and less noticing of what is being set aside.

How does all of this connect with complexity? Complexity themes that are implicit in my story include self-organizing (e.g., a Bronx zoo pathologist wires together a network of laboratories that make accelerated progress in isolating the virus), emergence (e.g., interactions generate an error and then a recovery from error), nonlinearity (e.g., unreturned phone calls trigger an effective workaround that solves the problem), semi-independent agents (e.g., conference calls are restricted to a small subset of players), dynamic unfolding (e.g., later results recast the meaning of earlier results), and turbulence in the form of overload, interruptions, understaffing, media pressure, unanticipated surge.

---

[12] The degree to which contributing and representing and subordinating are heedful, within a network of laboratories may influence the accuracy of diagnoses. For example, the initial reluctance to examine dead crows can be interpreted as a lack of subordination to and recognition of the system that is working the problem, since birds are reservoirs for vector-borne viruses, bird deaths are sentinels of larger human problems, and since human serum, which is what they were analyzing for NYC, was half of what pathologist Tracey McNamara wanted CDC to examine.

[13] The 'fallacy of centrality,' first described by Ron Westrum and elaborated by Weick (1995) and Snook (2000:173) consists of this logic: If there was viral traffic from the Old World to the New World, I would know about it. I don't know about it. Therefore, it isn't going on...and we can stop thinking about it! The fallacy is the belief that I am at the center of the flow of information. If crows and people were dying of the same thing, I'd know about it. CDC initially conducts a general, generic footprint test that is "adequate" to detect things they *do* know about. (Gill 2000:9,11). It is conceivable that even though CDC says publicly that its biggest challenge is to "be prepared to expect the unexpected," privately they believe that the unexpected is extremely rare, and that they know most of what there is to know.

When I talked about these complexity themes, I replaced complexity concepts with concepts from cognition, sensemaking, workflow interdependence, and interrelating. My argument is that these substitutions retain the spirit of complexity analysis but customize those insights so that they better fit human organization. Examples of these substitutions include,

In place of *CAS*, I talked about retrospective sensemaking. Agents pay attention to such things as open-ended acting, meaning as emergent, small actions that can have large effects, and surprise as an outcome.

In place of *Unknowability* I talked about partial connections that produce multiple realities; plausibility supported by justification rather than accurate representation; uncertainty as an issue of ontology rather than an issue of epistemology.

In place of *Partial* connections I talked about distributed sensemaking, semi-independent agents, reciprocal reference, identities that hold agents together, loosely coupled systems, and self-organizing fragments whose significance does not lie in the fact they there were once part of a greater whole.

In place of *Chaos* I talked about ambivalence, equivocality, ambiguity, and the unexpected.

In place of *Emergence* I talked about becoming, organizing, and juxtapositions that force novel meaning.

In place of *Dynamic* I talked about fluid, impermanent, process, ongoing, updating, exploration.

In place of *Co-evolution* I talked about reciprocal enactment of both the organization and the environment.

In place of *Self-organizing* I talked about organization that emerges *in* communication.

In place of *Simple* rules applied locally I talked about micro states that are central in organizing.

In place of *Non-linear* I talked about deviation amplifying feedback and small actions that can have large consequences (e.g., an unreturned phone call).

In place of *Entropy* I talked about normalizing, codification, shareability constraints, labeling.

In place of *Diversity* I talked about requisite variety, conflict, multiple drafts.

The resulting picture suggests that, as people connect and represent their joint contributions, more heedfully, they are more likely to differentiate and refine existing categories, create new categories, and perceive and enact a more nuanced context. When heedful interrelating produces mindful action, this is an example of people acting in order to think. What is different is that the acting is more relationally sensitive and the thinking is more situationally mindful. With fuller attention there is less confirmation bias. This line of argument is an example of the more basic point that different forms of task-based interdependence among players seem to induce different ways of thinking. In other words, "the cognitive properties of human groups may depend on the social organization of individual cognitive capabilities" (Hutchins 1995:176). To prepare for the unexpected means that you have to offset strong cognitive predispositions such as confirmation bias, fallacy of centrality, hubris, normalization, typification, and bottom-up salience of cues.

This is a complex map of interorganizing, but then the territory of everyday distributed sensemaking that it maps is no less complex. What matters is whether the map is useful. At this stage, I think it is. The model suggests that people misdiagnosed the West Nile Virus because of the way they were organized. Flawed interrelating enacted a flawed collective mind that was prone to lock-in familiar interpretations until interrelating became more heedful.

And what did all of this boil down to? What was the takeaway of this entire incident for CDC? The answer is a perfect example of non-linear relations. The West Nile incident taught people to return their phone calls!

References

Allinson RE (1993) Global disasters: Inquiries into management ethics. Prentice Hall, New York

Anderson P (1999) Complexity theory and organization science. Organization Science 10(3):216-232

Asnis DS, Conetta R, Teixeira AA, Waldman G, Sampson, BA (2000) The West Nile Virus outbreak of 1999 in New York: The Flushing Hospital experience. Clinical Infectious Diseases 30:413-418

Boyle RH (2000) Flying fever. Audubon 102 (14) Social psychology. Guilford, New York

Cook RI, Woods DD (1994) Operating at the sharp end: The complexity of human error. In: Bogner MS (ed) Human error in medicine. Erlbaum, Hillsdale, pp 255-310

Despommier D (2001) West Nile story. Apple Tree Productions, New York

Drexler M (2002) Secret agents: The menace of emerging infections. Joseph Henry Press, Washington DC

Eisenberg E (1990) Jamming! Transcendence through organizing. Communication Research 17(2):139-164

Eisenhardt KM, Bhatia MM (2002) Organizational complexity and computation. In: Baum JAC (ed) The Blackwell companion to organizations. Blackwell, Oxford, pp 442-466

Flin RH (1996) Sitting in the hot seat: Leaders and teams for critical incident management. John Wiley, New York

GAO (General Accounting Office) (2000) West Nile Virus outbreak: Lessons for public health preparedness (Report No. GAO/HEHS-00-180). Washington DC: General Accounting Office

Gill JM (2000) Expect the unexpected: The West Nile Virus wake up call, Report to Senator Joseph I. Lieberman. Washington DC: Minority Staff of Senate Governmental Affairs Committee

Hall SS (2003) On the trail of the West Nile Virus. Smithsonian, 14(4):88-102

Henig RM (1993) A dancing matrix: How science confronts emerging viruses. Vintage, New York

Hutchins E (1995) Cognition in the wild. MIT Press, Cambridge

James W (1992) Writings 1878-1899. The Library of America, New York

Last JM (ed) (2001) A dictionary of epidemiology, 4th edn. Oxford, New York

Mandler G (1984) Mind and body: Psychology of emotion and stress. Norton, New York

Nicholson N (1995) Enactment. In: Nicholson N (ed) Blackwell Encyclopedic Dictionary of Organizational Behavior. Blackwell, Cambridge, pp 155-156

Paget M (1988) The unity of mistakes: A phenomenological interpretation of medical work. Temple University, Philadelphia

Pfeffer J, Salancik GR (1977) Organizational design: The case for a coalition model of organizations. Organizational Dynamics, Autumn, 6(2):15

Rasmussen J (1983) Skills, rules, and knowledge: Signals, signs and symbols, and other distinctions in human performance models. IEEE Transactions on Systems, Man and Cybernetics 13(3):257-266

Scott E (2002) The West Nile Virus outbreak in New York City (A): Case C16-02-1645.0. Harvard University, Kennedy School of Government, Boston

Snook S (2000) Friendly fire. Princeton University, Princeton

Steinhauer J (1999, October 16) Battles over turf in health arena: Response to a viral outbreak highlights city-state tension. New York Times, Section B, p 1

Steinhauer J, Miller J (1999, October 11). In New York outbreak, glimpse of gaps in biological defenses. New York Times, Section A, p 1

Taylor JR, Van Every EJ (2000) The emergent organization: Communication as its site and surface. Erlbaum, Mahwah

Thompson JD (1967) Organizations in action. McGraw-Hill, New York

Vaughan D (1996) The challenger launch decision: Risky technology, culture and deviance at NASA. University of Chicago, Chicago

Wadler J (1999, October 1) Passionate life in a lab with dead animals. New York Times, Section B, p 2

Weick K E (1995) Sensemaking in organizations. Sage, Thousand Oaks

Weick KE, Roberts KH (1993) Collective mind in organizations: Heedful interrelating on flight decks. Administrative Science Quarterly 38:357-381

White DJ, Morse DL (eds) (2001) West Nile Virus: Detection, surveillance and control. The New York Academy of Sciences, New York

Section III Differing Views of Uncertainty and Surprise

IIIA Fundamental Unknowability in Science and Social Science

6 Fundamental "Uncertainty" in Science

Linda E. Reichl[1]

The conference on "Uncertainty and Surprise" was concerned with our fundamental inability to predict future events. How can we restructure organizations to effectively function in an uncertain environment? One concern is that many large complex organizations are built on mechanical models, but mechanical models cannot always respond well to "surprises." An underlying assumption about mechanical models is that, if we give them enough information about the world, they will know the future accurately enough that there will be few or no surprises. The assumption is that the future is basically predictable and *deterministic.*

The concept of determinism arose in the 18th century after the great success of Newtonian mechanics. In 1686 Newton published the first of his three volumes called the *Principia.* In those books, he gave his three laws of mechanics: 1) a body maintains its state of rest or uniform velocity unless acted on by a force, 2) **F** = m**a** (Force = Mass times Acceleration), and 3) for every action there is an equal and opposite reaction. Newton's *Principia* contained the mathematics necessary to explore the implications of these three laws. In addition, in the *Principia,* Newton proposed his inverse square law of gravitation and used it to 1) derive Kepler's empirical laws of planetary motion, 2) explain the motion of the moon and the tides, and 3) explain the behavior of falling bodies. Newton's *Principia* marked the beginning of modern science. It is not surprising, in view of the great success of Newtonian mechanics, that many individuals transferred their hope of predicting the future from religion to science. It gave rise to the philosophy of determinism, which is perhaps most simply stated by Laplace: *"If you give me the positions and momenta of all the particles in the Universe, I will predict all past and future."* This deterministic view of the world is still held by many scientists to this day.

On April 8, 1986, in a lecture to the Royal Society on the 300th anniversary of Newton's *Principia,* Sir James Lighthill made the following statement:

> I speak...once again on behalf of the broad global fraternity of practitioners of mechanics. We are all deeply conscious today that the enthusiasm of our forbears for the marvelous achievements of Newtonian mechanics led them to make generalizations in this area of predictability which...we tended to believe before 1960, but which we now recognize as false. We collectively wish to apologize for having misled the general educated public by spreading ideas about determinism of systems satisfying Newton's laws of motion that, after 1960, were proven to be incorrect. (1986)

Newton's laws, except for a countable number of cases, cannot determine the future exactly, they can only give the probability that certain events will happen.

---

[1] Professor of Physics and Acting Director, Center for Studies in Statistical Mechanics and Complex Systems, The University of Texas at Austin, reichl@physics.utexas.edu.

There are actually only a few examples of systems for which Newtonian mechanics can predict the future for very long times (Reichl 1992). Most conservative (energy conserving) Newtonian systems are chaotic, and for chaotic systems one can only predict the future for a very short time because any uncertainties in initial conditions or lack of accuracy of input data (such as the value of Pi) will grow exponentially with time. After a short time interval, we must resort to probabilistic descriptions of the future evolution of the system.

One of the goals of science over the past several thousand years, and a primary driving force in the creation of the *Principia*, has been to understand the future behavior of the solar system. This, of course, is critical for our survival. It happens that the solar system is one of the most regular mechanical systems known. The solar system is a conservative Newtonian dynamical system (neglecting space junk and other small effects) consisting of the nine planets and the sun, held together by Newton's gravitational force which acts between the planets and the sun, and between the various planets. The solar system does not exhibit hard chaos. In fact, on short time scales (of order millions of years) its motion is very regular. However, recently it was shown (Laskar 1996) that on longer time scales the eccentricity of the orbits of the inner planets, Mercury, Venus, Earth and Mars have a chaotic component. The eccentricity of the Mercury's orbit, in particular, can undergo large chaotic excursions which can cause it to intersect the regions of excursion of Venus's orbit. Thus, their long time future motion (hundreds of millions of years) cannot be predicted accurately. There is a small chance that they could collide, but because the motion is chaotic we can not know if this will actually happen. We only know that Newtonian mechanics predicts that it might happen.

What is it that causes mechanical systems to undergo a transition to chaos? The answer is *resonance* between the various degrees of freedom of the system. Resonances cause efficient transfer of energy. Resonance occurs when fundamental periods of oscillation of the various degrees of freedom become commensurate. Any given degree of freedom may have several different periods associated with its motion, describing different time scales. If the system is nonlinear (not a harmonic oscillator) these periods can change as the energy of the system changes. Generally when resonance occurs, it occurs on many different length scales and time scales, simultaneously, and self-similar resonance networks form in the phase space of the system. When the resonances overlap and intersect, you get chaos. Because of the efficient interchange of energy at many different length scales and time scales, the system can explore all the states available to it. When this chaotic behavior sets in, you can only predict the future behavior of the system in a probabilistic manner.

Even though it may not be possible to predict the future of chaotic systems with any certainty, one thing is completely certain - the stuff the world is made of will not disappear. This is so because the physical world has some underlying symmetries. One of the most important symmetries occurs in systems for which the Newtonian mechanics is time translation invariant. Time translation invariance gives rise to the conservation of energy[2] (Reichl 1992). Because energy cannot be destroyed, the world can exist. We can have galaxies. In fact, a galaxy is just a huge bundle of energy which is diversified into many different forms. We don't yet understand all the different forms of energy that exist in the universe. We don't yet understand the dynamics of the universe. Scientists are now trying to understand recent surprising observations concerning the expansion of the universe, in terms of something called "dark matter" which has not yet been measured. It is all an open question right now. But since the focus of this conference is the dynamics of the complex structures on planet Earth, let us return there.

Planet Earth, and everything on it, exists in its present form because of conservation of energy and the lucky placement of the Earth in a solar system with fairly regular dynamics. Because of the comparatively gentle environment the Earth finds itself in the solar system, a vast network of complex chemical and biological structures has formed on the surface of planet Earth. In his keynote lecture, Peter Allen talked about something called the "Brusselator," which is a simple chemical system that is the prototype of many complex systems (Nicolis and Prigogine 1977; Reichl 1998). The Brusselator model helps explain how these complex structures on Earth could have formed. I would like to try to place Brusselator dynamics into the perspective of the Newtonian dynamics described above. The Brusselator describes the complex (nonlinear) non-equilibrium dynamics of a system containing a trillion trillion molecules which can interact and react, and by reacting they can change their structure. The underlying microscopic dynamics of these trillion trillion molecules is totally chaotic. On the macroscopic scale, it is a thermodynamic system and thermodynamics is founded on chaos. You can imagine a container of these molecules which are continually undergoing catastrophic collisions. They are breaking apart and recombining in a random manner. If you could visualize the huge phase space of these trillion trillion molecules, it would be totally chaotic. During the collisions between molecules, everything that is not conserved due to some underlying symmetry is destroyed. Indeed, the only thing conserved is the energy. Because the energy of motion of the molecules is non-relativistic, the congealed energy which forms the mass of the molecules cannot be changed to other forms. Thus, the total mass of the molecules doesn't change although it can get redistributed. Because mass is conserved, you can count the molecules, and that ability to count molecules gives rise to the only quantity available to describe the system on a thermodynamic scale - the density of the various molecules (the density is the number of particles per unit volume of the system).

If you let such a system evolve in a test tube, it will approach an *equilibrium* state where the density of various types of molecules becomes uniform throughout the system. Even though the molecules are continually colliding and interacting, in any given time interval, as many molecules of type "A" will be created as de-

[2] Strictly speaking in order to understand why the world does not disappear, one must deal with these concepts at the level of relativistic dynamics and quantum field theory, and even there the picture is not yet complete. It is sufficient for our purposes to consider non-relativistic dynamical systems.

stroyed and there will be no net change in their density. When the world reaches equilibrium, no complex *dynamical* structures can exist. The entropy (degree of disorder) of equilibrium systems is maximum.

Complex dynamical structures can only exist in a world held out of equilibrium. The Brusselator provides a model for how this works. The Brusselator refers to a chemical system with six chemical constituents, A, B, X, Y, D, and E. The chemicals A and B continually flow into the system. When A enters the system it undergoes a reaction and transforms into X. When B enters, it interacts with X to form Y and D. Then two molecules of X interact with Y to form three molecules of X. Finally X changes to E. These reactions may be written in short hand form as A □ X, B + X □ Y+ D, 2X + Y□3X, X □E. The chemicals D and E are pulled out of the system as soon as they form. The chemicals X and Y carry the interesting dynamics of the system. You have a continual flow of chemicals through the system, thus forcing it to remain far from thermodynamic equilibrium.

You can write equations of motion for the densities of the various chemicals in the Brusselator (Nicolis and Prigogine 1977; Reichl 1998) (these are called reaction-diffusion equations) and these equations can be solved numerically to investigate the dynamics. It is found that, in this far-from-equilibrium chemical system, chemical structures form. The two chemicals of primary interest are X and Y. Under certain conditions, the densities of X and Y form spatial and temporal patterns. For example, they can form chemical clocks where at one time mostly X exists in the system and a few seconds later mostly Y exists. This oscillation in densities of X and Y can persist as long as chemicals are fed into the system. Also spatial patterns can form in which regions containing large quantities of X might alternate in space with regions containing large quantities of Y. Chemical waves have also been observed. The key to the formation of these chemical structures is the nonlinear nature of the chemical reaction. This appears in the form of a feedback loop in the reaction scheme – it requires two molecules of X to create three molecules of X (2X+Y□3X). This creates a non-linearity in the equations of motion of this reaction. The Brusselator reaction has been realized in the laboratory using a chorite-malonic acid reaction in a two-dimensional gel (Reichl 1998).

The Brusselator is the simplest chemical model which exhibits non-equilibrium chemical structures. A more complicated example is the Belousov-Zhabotinsky (B-Z) reaction (Nicolis and Prigogine 1977; Reichl 1998). The Belousov-Zhabotinsky reaction involves a much more complicated set of chemical reactions, but its essential features can be reproduced with three key types of molecules. This reaction also has feedback loops which make it nonlinear and, when it is run far from equilibrium by allowing a continual flow of chemicals, one also sees chemical clocks, spatial variations in the concentrations, and chemical waves. Similar types of nonlinear chemical reactions and chemical structures have been observed in biological systems and are thought to be essential to maintenance of life.

At thermodynamic equilibrium, dynamic chemical structures, like those observed in the Brusselator and B-Z reactions, cannot exist. In order to create them, the system must be driven far from chemical equilibrium, by continually adding and removing certain chemicals. Also the structures don't emerge gradually. They emerge suddenly as one moves far enough away from equilibrium. When they were first observed, they were a complete surprise. One must remember that on the microscopic scale the system is completely chaotic and everything that can be destroyed will be destroyed. Only the density cannot be destroyed. And yet out to this microscopically chaotic system, order emerges on macroscopic space and time scales (as long as the macroscopic system is sufficiently far from thermodynamic equilibrium). The entropy of these systems is lower than their equilibrium counterparts, although they continually produce entropy in the world around them. They feed on low entropy ordered systems and expel higher entropy by-products.

Of course, this conference is about systems composed of living beings, like buffalo or people, and not molecules. How do we understand the dynamics of complex structures created by and composed of complex living beings? Often the chemical systems described above have been used to understand some features of complex social systems.

Complex social systems are equally fragile. They require a continual flow of energy to maintain their existence, and they often require a continual flow of information, the analog of entropy. If the energy flow stops, they disappear. Complex social systems can suddenly change their structure and change into something completely new and unpredicted as some key parameters are changed. They are often best sustained if they have a very efficient communication network, which involves flow of information throughout the entire structure.

Of course, one must be careful in making comparisons between chemical systems and complex social systems. To be sure, there are many lessons to be learned by studying the chemical systems. But there are also significant differences, and the effect of these differences is not understood. In systems composed of living beings, the constituents themselves store energy and information. They can learn and change their behavior according to what they have learned. They can also create new structures in response to what they see as the needs of their surroundings. Another important difference is that living beings (unlike molecules) are not conserved. They live and function for a finite time and then are gone. There is no conservation law at work here. From the point of view of long time dynamics, this is perhaps the more important and fundamental difference between chemical systems and social systems. Which brings us to time scales governing chemical and social systems: The interaction times between chemical constituents can be extremely short, of the order of 10^{-15} seconds, and yet the chemical clocks observed on the macroscopic level oscillate on times scales of the order of seconds. Thus there is a huge difference between the microscopic and macroscopic time scales of interest in chemical systems.

What about the dynamics of social systems? Does the ability of the constituents to learn change the time scales necessary to change the system as a whole? Obviously it does and one must be careful in following the chemical analogies too closely. There is much more synergy between the dynamics of the constituents and the dynamics of the entire complex social system. But still, the amazing structures that can form suddenly in complex chemical systems, when they are changed slightly, must give us pause when we contemplate changing even some small aspects of complex social systems.

References

Laskar J (1996) Large scale chaos and marginal stability in the solar system. Celestial Mechanics and Dynamical Astronomy 64:115-162

Lighthill J (1986) The recently recognized failure of predictability in Newtonian dynamics. Proc. Roy. Soc. A407:35-48

Nicolis G, Prigogine I (1977) Self-organization in nonequilibrium systems. Wiley-Interscience, New York

Reichl LE (1998) A modern course in statistical physics. Wiley-Interscience, New York

Reichl LE (1992) The transition to chaos in conservative classical systems: Quantum manifestations. Springer, New York

7 The Complementary Nature of Coordination Dynamics: Toward a Science of the In-Between

J.A. Scott Kelso[1]

7.1 Introduction: Whither a Science of the In-Between?

My take-off point in this brief essay is the following comment from the introductory material that all the participants received for the conference on "Uncertainty and Surprise." It reads:

> *Instead of* viewing social systems as machines whose Newtonian-like dynamics are to be uncovered and then controlled, we now see social systems as self-organizing systems whose properties emerge from interactions between *agents.* (italics mine)

One is, of course, sympathetic to this view in general (Kelso 1995, 2001a). At the same time, however, it does not seem necessary to cast the benefits of self-organizing dynamics as an alternative to Newtonian mechanics in an either/or fashion. As James Gleick (2003) remarks in his recent book *Isaac Newton* (see also the even more scholarly *Never at Rest* by Richard Westfall), Newton himself never succumbed to the fantasy of pure order and perfect determinism. Newton already saw that chaos could emerge in the interactions of many bodies. "Unless I am much mistaken" Newton said, "it would exceed the force of human wit to consider so many causes of motion at the same time, and to define the motions by exact laws which would allow of an easy calculation" (cited in Gleick 2003:137). Any developments in self-organizing dynamical systems rest on the shoulders of Newton, or at least they have his fingerprints all over them. We do not need to throw out determinism merely because we embrace uncertainty (consider, to boot, the huge field of study dealing with *deterministic* chaos). The main general point here concerns the habit we have of putting things in terms of dichotomies. More specifically, I believe we need a science that embraces not only the extremes, but also the vast world of the in-between (Kelso and Engstrom 2005). That science is emerging and has gathered a good deal of impetus in the last 25 years or so. In the literature, it has a name: it is called *coordination dynamics.* The term *agents* also italicized in the quotation at the beginning of this chapter, and the issue of how agency arises from and directs or steers self-organizing dynamics is addressed elsewhere (Kelso 2002; Kelso and Engstrom 2005).

---

[1] Complex Systems and Brain Sciences, Florida Atlantic University, Boca Raton, kelso@ccs.fau.edu.

Ask yourself: Why do we talk about "instead of" and "versus" all the time? Why do we partition the world into pairs, contrasting, for example, the genotype and the phenotype, the discrete and the continuous, the individual and the collective, the orderly and the random, the qualitative and the quantitative, the internal and the external, the persistent and the changing, the gradual and the abrupt, the reductionist and the holist - and yes, the certain and the uncertain. The list goes on and on. It is pretty obvious, one intuits, that both represent polarized extremes, and that reality must lie somewhere in between. One might even say that we categorize things and ideas in this polarized fashion in order to be sure that what we are really after will be captured in between. One might even say that all science is about the in-between. One might, indeed. Few have.

Such thinking raises several questions. Is it something about our own brains that makes us categorize things in either/or terms? If so, how do we understand that? And what if we were to view things from the perspective of "both/and" rather than "either/or" (see how insidious is the habit to dichotomize!). Let me say that again: What if we were to embrace *both* the either/or and the both/and, *and* everything in between? What would such a science look like? At the very least, such a science, I submit (e.g., Kelso 1995, Chap. 4; Kelso and Engstrom 2005) would have to include *both* the language of states, in which polarized extremes may be seen as stable states (stationary attractors) of a dynamical system, *and* the language of tendencies or dispositions, in which there are no states (stable or unstable) at all. In fact (and I use the word "fact" carefully), the science of the in-between - like a James Joyce narrative - consists *of multiple tendencies coexisting at the same time.* The science of the in-between - and importantly, the philosophy that motivates and accompanies it - thus represents a strange kind of complementarity (Bohr 1935). We call it The Complementary Nature (Kelso and Engstrom 2005).

On first blush, all this may seem rather obscure, philosophical and speculative, so let's ground it in an example. In fact, let's take a nontrivial example, one that lies close to the very issue of why we might split the world into pairs in the first place. The example concerns our understanding of the brain and - when the brain is embedded in its world - the behavior it gives rise to.

Historically, there are two (!) main theories of how the brain works (see Finger 1994 for an excellent historical account). One is that the brain works as a coordinated, integrated entity. The other is that it is segregated into highly specialized parts that are localized for particular functions. Why, a reasonable person asks, could the brain not be both? We could all agree, "Of course! Let's have both." Instead of viewing integration and segregation as conflicting processes (and theories), let's view them as complementary. That sounds grand, but what does it really look like? What would be the manifestation of having both segregation and integration in the brain, or for that matter in any naturally complex system like a society? (Ask your favorite politician). In what way might integration and segregation be construed and understood as complementary?

To come to grips with these questions, we have to take a short sojourn into coordination dynamics, a theory of how coordinated patterns of behavior arise in a self-organized fashion, and how they adapt, persist and change according to internal and external circumstances (see Jirsa and Kelso 2004 for a sampling of recent work in this field). Coordination dynamics aims to characterize the nature of the coupling *within* a part of a system, *between* different parts of a system and between *different kinds of systems*. Moreover, it explicitly addresses the connection between different levels of description (see e.g., Kelso et al. 1999). Ultimately, coordination dynamics is concerned with how things come together in space and time, and how they break apart - bringing us back full circle to the integration~segregation issue.

From now on, by the way, we'll use the squiggle or tilde (~) as part of a convenient syntax for complementary pairs (Kelso and Engstrom 2005). The squiggle does not represent a glue or a bridge. Rather it is to acknowledge, in a world replete with either/or dichotomies, that complementary aspects are inextricably related, yet each may retain their singular character.

7.2 Coordination Dynamics

Consider the large scale spatiotemporal dynamics of the brain. Measures of brain activity using current imaging techniques such as EEG, MEG and MRI all depend on synchronous activity in populations of neurons, so-called neural ensembles. Without the basic cooperative effect of synchronization among approximately 10^5 neurons, no signal would even be observable. Synchronization, of course, is a classic example of self-organized coordination (Kelso 1995; Haken 1996 for reviews) and may be described by a model of N globally-coupled nonlinear oscillators (Kuramoto 1984), here the neurons themselves:

$$\frac{d\phi_k}{dt} = \omega_k + \frac{K}{N}\sum_{j=1}^{M} \sin(\phi_j - \phi_k). \tag{7.1}$$

Eq. (7.1) describes the evolution in time of the oscillator phases ϕ_k. A critical coupling parameter K determines different modes of synchronization, and is a function of the dispersion of individual oscillator frequencies, ω_k.

Segregation means that neural ensembles in different regions of the brain fire independently of each other. Integration means that there is some kind of coordination between different regions - often across large distances - thereby implying patterns of functional connectivity that evolve in time (Sporns and Tononi 2002). Without going into details, there are both experiments (e.g., Daffertshofer et al. 2000; Fuchs et al. 2000; Kelso et al. 1992, 1998) and theory (e.g., Haken, Kelso and Bunz 1985; Jirsa et al. 1998; Kelso et al. 1990) that show that the brain's coordination dynamics may take the form of transient phase-locking within and between different neural regions (see Varela et al. 2001 for review). The basic idea is that phase-locked oscillations in different brain areas such as the cerebellum, hippocampus, thalamus, olfactory cortex and neocortex can serve a "binding" function. For example, perception may arise as a result of temporary episodes of phase-locking at the γ-frequency (approx. 40 Hz) in the brain thereby linking stimulus features into a coherent, coordinated Gesthalt. Phase-locking synchrony appears to be a universal process of communication that transcends several levels of neural information processing. There is even evidence that individual neural spikes encode information about the synchronization process, a kind of basic "temporal coding" (Ermentrout and Kopell 1998).

Given that different regions in the brain may fire at different frequencies (there are Greek names for them, alpha, beta, delta, mu and theta, all tied to various functions such as sensorimotor processing, remembering, attending, and so forth) and given that the brain has mechanisms for regulating these frequencies (which it does) the synaptic connectivity among neural pools is captured best by the relative phase, ϕ_{ij} between the *j*th and *i*th oscillating regions. Here again, many experiments have shown that different phase relations are possible among interacting brain areas and that abrupt transitions may occur as relevant parameters are varied (e.g., Mayville et al. 2000; Meyer-Lindenberg et al. 2002).

A simple, but essentially nonlinear dynamics that captures coordinative stability and switching between coordinative states in both brain and behavior is:

$$\dot{\phi} = -a\sin\phi - 2b\sin 2\phi, \tag{7.2}$$

where ϕ is the relative phase or natural phase difference between the component parts, $\dot{\phi}$ is the derivative of ϕ with respect to time, and *a* and *b* are control parameters, the ratio of which (*b*/*a*) specifies the coupling among the interacting components. An equivalent formulation of Eq. (7.2) is

$$\begin{aligned} \dot{\phi} &= -\partial V(\phi)/\partial\phi \text{ and} \\ V(\phi) &= -a\cos\phi - b\cos 2\phi. \end{aligned} \tag{7.3}$$

In the literature, this is called the HKB model of coordination, after Haken, Kelso and Bunz (1985) who formulated it as an explanation of observed multistability, phase transitions and hysteresis in the coordinated movements of human beings (Kelso 1984). For the record, Eqs. (7.2) and (7.3) have two stable *fixed point attractors*, corresponding to coordinative phase-locked states at $\phi = 0$ and $\phi = \pm\pi$ rad. Thus, two (and, in general, multiple) coordinated states may *coexist* for exactly the same parameter values, the essentially nonlinear feature of *bistability*. Which one is observed, of course, depends on initial conditions that set the system into a particular basin of attraction. As the coupling (*b*/*a*) in Eqs. (7.1) and (7.2) varies, the coordination states at $\phi = \pm\pi$ rad. become unstable and switch - fully in line with experiments at both behavioral and brain levels (op. cit.).

Very many extensions and elaborations of this elementary coordination dynamics have occurred in the last 20 years (see Jirsa and Kelso 2004), but there is one key one that will turn out to be crucially important for seeing the connection between the language of states and the language of coexisting tendencies. It concerns *broken symmetry*. Notice that Eqs. (7.2) and (7.3) are *symmetric*: the coordination dynamics is 2π periodic and is identical under left-right reflection (ϕ is the same as minus ϕ). This means that in-phase and anti-phase belong to the same (spatio-temporal) symmetry group, and assumes that the individual components are *identical*, a condition in living things that is seldom, if ever, satisfied. As already pointed out, neural oscillators are divided into pools or ensembles according to their different natural frequencies. In general, nature thrives on *broken symmetry*.

To accommodate observations of broken symmetry, Kelso et al. (1990) introduced a term $\Delta\omega$ into the dynamics as follows:

$$\begin{aligned} \dot{\phi} &= \Delta\omega - a\sin\phi - 2b\sin 2\phi \text{ and} \\ V(\phi) &= -\Delta\omega\phi - a\cos\phi - b\cos 2\phi \end{aligned} \tag{7.4}$$

for the equation of motion and the potential respectively. $\Delta\omega$ represents intrinsic differences between the components, i.e., in their natural frequencies, and attests to the heterogeneity of the interacting components. Small values of $\Delta\omega$ shift the attractive fixed points of the coordination dynamics in an adaptive manner, adjusting system behavior to the intrinsic properties of (and differences between) the individual components For larger values of $\Delta\omega$ the attractors disappear: no coordination between the components occurs. That is, the individual parts behave as separate, autonomous entities.

There is, however, an enormously interesting régime *in between* the idealized states of full cooperation and the total independence of the component parts from each other. This is the *metastable régime* in which there are no attractors or repellors (fixed points of the coordination dynamics) but there is still *attraction* to where the attractors used to be (Kelso 1991). The reason is that intrinsic differences ($\Delta\omega$) between the individual component parts are sufficiently large that they do their own thing, while the coupling is sufficient to hold the parts together so that they still retain a *tendency* to cooperate. Thus, the relative phase between the components drifts, but is occasionally trapped near remnants or ghosts of where the coordinated states used to be (i.e., near $\phi = \pm\pi$ and $\phi = 0$). This, I propose, is how global integration, in which the component parts tend to be locked together in harmony ("binding"), is reconciled with the tendency of the parts to function locally as specialized autonomous units. Because of broken symmetry in its coordination dynamics, the brain is able to exhibit a far more variable, plastic and fluid form of coordination in which integration and segregation tendencies *coexist at the same time,* metastable coordination. Metastable coordination dynamics is characteristic of successful organizations, and is especially evident in the human brain (Kelso 1995; see Bressler and Kelso 2001 for a recent review). Elsewhere, I have argued, this is where a certain kind of flexibility and stability (metastability) gives rise to the creation of information (Kelso 2001b).

7.3 Discussion

Gathering these results together and returning to the business at hand, we see two extremes in coordination dynamics - even in its most elementary form. One is for the component parts to be coupled together in coherent (self-organized) coordinated states. These are stable states: stationary attractors of a dynamical system. These stable states correspond to "pure" integration. The other extreme is for the component parts to be completely independent, doing their own thing. There is no coordination or interaction among the parts whatsoever, total segregation. Only two (!) factors are running the show here: the strength of the coupling between the parts (*b*/*a* in Eqs. (7.2), (7.3), and (7.4)) and the intrinsic differences between the components, a complementary pair consisting of coupling~components.

In between, however, is the huge territory of tendencies and dispositions (In this regard, one is reminded of the Oxford philosopher, Gilbert Ryle, who, in his 1949 book *The Concept of Mind*, considered beliefs, intentions, desires as dispositions toward behavior depending on context.). Notice that this "in between" is chock full of complementary pairs: Individual component parts coexist with the collective (individual~collective); cooperation coexists with competition (competition~cooperation) - the component parts competing with each other in order to retain their autonomy while also trying to cooperate; in the flow of the dynamics, the tendency to converge toward attractive fixed points (phase-locked states) and the tendency to diverge to (coexisting) repelling fixed points (convergence~divergence) (see Kelso 1995:104-114 for pictures of this); qualitative change (phase transitions) produced by quantitative variation of parameters, and accompanied by quantitative consequences such as enhancement of fluctuations (qualitative~quantitative); and, of course, the tendency to integrate coexisting with the tendency to segregate, thus allowing us to see - through the window of coordination dynamics - how we might in fact have a science that embraces both the both/and *and* the either/or. In other words: A science that undergirds both polarization and reconciliation.

Let me finish this brief essay with just a few more points (from a very wide range of possibilities). The first concerns our notion of complementary pairs, which is as old as the days and goes back at least to the Ionian philosopher Heraclitus (540 B.C.-480 B.C.). Heraclitus, the reader will remember, said the world would not exist without the clash of opposing currents. This theme of opposing "contrarities" (Samuel Beckett's lovely word, or is it Joyce's?) runs through the history of ideas and the men and women who generated them (Plato, Aristotle, Aquinas, Kant, Hegel, Marx, De Beauvior....). Science too has struggled (and still struggles) with its complementary pairs and how to reconcile them (discrete and continuous, space and time, mass and energy, electricity and magnetism, gradualism and saltationism, nature and nurture, genotype and phenotype, central controls and laissez-faire, individual and collective, etc., etc.).

In more modern times, the notion of contrasting opposites is central in the writings of the theoretical physicist-mystic, Fritjof Capra (2000):

The notion that all opposites are polar - that light and dark, winning and losing, good and evil are merely different aspects of the same phenomenon - is one of the basic principles of the eastern way of life. Since all opposites are interdependent their conflict can never result in the total victory of one side, but will always be a manifestation of the interplay between two sides... (p. 146)

Capra goes on to say that the notion of dynamic balance is essential. It is never a static entity but always a dynamic interplay between two extremes: "The dynamic unity of opposites can be illustrated with the simple example of a circular motion and its projection (onto a line, my words)....The circular movement will appear as an oscillation between two opposite points, but in the movement itself the opposites are unified and transcended..." (Capra 2000:147).

This is a pretty image of Capra's - the Tao is in the middle of the circle and the two extremes of the line are yin and yang. As the ball goes around in the circle it moves up and down on the line. From this (metaphorical) image of the dynamic unification of opposites, Capra proceeds to the microscopic realm. His reason is because "our classical notions derived from our ordinary macroscopic experience are not fully adequate to describe this subatomic world [which] appears as a web of relations between the various parts of a unified whole" (p. 159).

"Contrarities" are all around. They are ubiquitous. Contrarities are complementary (as Bohr's coat of arms says: *Contraria sunt complementa*). But the scientific basis of complementary pairs and what it means for them to be dynamical is either metaphorical (the "dynamic" interplay of opposites) or rests on an interpretation of how the subatomic world behaves, viz., the Copenhagen interpretation of Quantum Mechanics. Coordination dynamics lies - you guessed it - somewhere in between the classical world of Newtonian mechanics with its forces, masses and motions and the weird, but highly successful world of Quantum Mechanics with its probabilistic waves and particles. Here, in coordination dynamics - which deals in the currency of coordination variables like phases and amplitudes of brain and behavioral and social activity - we offer an explanation/interpretation of complementary pairs that is neither metaphorical nor (solely) quantum mechanical in origin. Coordination dynamics ties "polar opposites" like integration and segregation to essential nonlinearity (bistability in the simplest case) and their mutual interplay to coexisting tendencies that arise in the metastable regime of the dynamics. I find it amusing that Newton saw the self motion that God gave animals "beyond our understanding without doubt" (not so the motions of inanimate bodies, at least for him!), yet the coordinated movements of human beings, which most of us take for granted, happen to be the primary stimulus for the development of a science of coordination. For living things, it seems, we must move from a mechanics of motion to a dynamics of coordination.

A second and related point concerns the relationship between coordination dynamics and Quantum Mechanics. Elsewhere I have addressed this connection, in particular the metastability inherent in both that is necessary for the creation of information (Kelso 2001, 2002). We owe to Niels Bohr, of course, the great philosopher~scientist that he was, the Copenhagen interpretation of Quantum Mechanics - that a full description of light and matter relies on complementarity.

Complementarity, according to Bohr, means that the use of certain concepts in the description of nature automatically excludes the use of other concepts, which however, are equally necessary for a description of the phenomenon. Despite its subtleties, complementarity, as John Archibald Wheeler (1994) emphasizes, is battle tested. There is no going back on it. It says, for instance, that one cannot find the position of an electron and the momentum of an electron at the same place and time. The measurement of one has unpredictable consequences for the other. The interaction of the measuring device (the brain?) and the object being measured (the world?) is never nil.

The complementary nature of coordination dynamics - with its built-in, essential nonlinearity - is different, and perhaps just as strange as Quantum Mechanics. It says that two opposing tendencies are complementary and do coexist at the same time. Measures such as the distribution of dwell times (how long the system hangs around a given tendency before it leaves) and its complement (how long it escapes for, when it leaves) open up ways to address the strength of the coupling among components relative to their independence, their individual freedom. More generally, the simultaneous presence of convergence and divergence in the flow of the coordination dynamics attests to its truly non-stationary, transient nature. In my view, here is where the science of the "in-between," the science of tendencies and dispositions, puts the language of stationary attractors (chaotic and otherwise) and the notion of "brain states" in appropriate relief.

This work was supported by a National Institute of Mental Health Senior Scientist Award. It is dedicated to the memory of Ilya Prigogine and Agnesee Babloyantz.

References

Bohr NHD (1937) Causality and complementarity. Philosophy of Science 4:289-298

Capra F (2000) The tao of physics. Shambhala, Boston

Bressler SL, Kelso JAS (2001) Cortical coordination dynamics and cognition. Trends in Cognitive Sciences 5:26-36

Ermentrout GB, Kopell N (1998) Fine structure of neural spiking and synchronization in the presence of conduction delays. Proceedings of the National Academy of Sciences (USA) 95:1259-1264

Daffertshofer A, Peper CE, Beek PJ (2000) Power analysis of event-related encephalographic signals. Physics Letters A 266:290-302

Finger S (1994) Origins of neuroscience. Oxford University Press, New York, Oxford

Fuchs A, Mayville J, Cheyne D, Weinberg H, Deecke L, Kelso JAS (2000) Spatiotemporal analysis of neuromagnetic events underlying the emergence of coordinative instabilities. NeuroImage 12:71-84

Haken H (1996) Principles of brain functioning. Springer, Berlin

Haken H, Kelso JAS, Bunz H (1985) A theoretical model of phase transitions in human hand movements. Biological Cybernetics 51:347-356

Jirsa VK, Kelso JAS (eds) (2004) Coordination dynamics: Issues and trends. Vol 1. Springer Series in Understanding Complex Systems. Springer, Berlin, Heidelberg

Jirsa VK, Fuchs A, Kelso JAS (1998) Connecting cortical and behavioral dynamics: Bimanual coordination. Neural Computation 10:2019-2045

Kelso JAS (1984) Phase transitions and critical behavior in human bimanual coordination. American Journal of Physiology: Regulatory, Integrative and Comparative 15:R1000-R1004

Kelso JAS (1991) Behavioral and neural pattern generation: The concept of neurobehavioral dynamical system (NBDS). In: Koepchen HP (ed) Cardiorespiratory and motor coordination. Springer, Berlin

Kelso JAS (1995) Dynamic patterns: The self-organization of brain and behavior. MIT Press, Cambridge

Kelso JAS (2001a) Self-organizing dynamical systems. In: Smelser NJ, Baltes PB (eds in chief) International encyclopedia of social and behavioral sciences. Pergamon, Amsterdam

Kelso JAS (2001b) How the brain changes its mind: Metastable coordination dynamics. In: The emergence of the mind. Fondazione Carlo Erba, Milano, pp 93-101

Kelso JAS (2002) The complementary nature of coordination dynamics: Self-organization and agency. Nonlinear Phenomena in Complex Systems 5:364-371 (Special Issue in honor of Hermann Haken)

Kelso JAS, Engstrom DA (2005) The complementary nature. The MIT Press, Cambridge

Kelso JAS, Fuchs A, Jirsa VK (1999) Traversing scales of brain and behavioral organization. I. Concepts and experiments. In: Uhl C (ed) Analysis of neurophysiological brain functioning. Springer, Berlin, pp 73-89

Kelso JAS, DelColle J, Schöner G (1990) Action-perception as a pattern formation process. In: Jeanerod M (ed) Attention and performance XIII. Erlbaum, Hillsdale, pp 139-169

Kelso JAS, Bressler SL, Buchanan S, DeGuzman GC, Ding M, Fuchs A, Holroyd T (1992) A phase transition in human brain and behavior. Physics Letters A 169:134-144

Kelso JAS, Fuchs A, Holroyd T, Lancaster R, Cheyne D, Weinberg H (1998) Dynamic cortical activity in the human brain reveals motor equivalence. Nature 392:814-818

Kuramoto Y (1984) Chemical oscillations, waves and turbulence. Springer, Berlin

Mayville JM, Fuchs A, Ding M, Cheyne D, Deecke L, Kelso JAS (2001) Event-related changes in neuromagnetic activity associated with syncopation and synchronization tasks. Human Brain Mapping 14:65-80

Meyer-Lindenberg A, Ziemann U, Hajak G, Cohen L, Berman KF (2002) Transitions between dynamical states of differing stability in the human brain. Proceedings of the National Academy of Sciences (USA) 99:10948-10953

Sporns O, Tononi G (2002) Classes of network connectivity and dynamics. Complexity 7:28-38

Varela FJ, Lachaux JP, Rodriguez E, Martinerie J (2001) The brainweb: phase synchronization and large-scale integration. Nature Reviews Neuroscience 2:229-239

Wheeler JA (1994) At home in the universe. American Institute of Physics, Woodbury

8 The Tyranny of Many Dimensionless Constants: A Constraint on Knowability

Bruce J. West[1]

My talk today is concerned with one particular aspect of knowability having to do with the modeling and simulation of complex phenomena. This approach is based on dimensional analysis and I hope to show how dimensionless constants can be used to quantify the complexity of systems and indirectly show the difficulty associated with extracting information and knowledge from such phenomenon.

The validity of scientific inference is tested by experiment, and experiment is often preceded by prediction, or at least followed by postdiction. In science it is the ability to explain the outcome of experiment that ultimately determines the confidence we have in what we know, either by prediction or postdiction. If we do not have a model that allows us to explain, then we feel that what we think we know is of limited utility. In physics what we mean by a model is a mathematical description, and a prediction is manifest in the solution to a set of model equations describing how the systems evolves from a realizable initial state to a measurable final state. The model encapsulates what we know and what is knowable.

Of course there is more than one way to construct a model of reality. In medicine, for example, physicians often refer to an "animal model," by which they mean a pig, or a monkey, or other animal that shares some property with humans that they wish to test. This notion of model replaces the physical idea of a mathematical model with a homologue or homology, that is, a similarity that is based on a common ancestry and is a likeness short of an identity in structure or function in different organisms.

Thus, one phenomenon that we do not understand and cannot control is replaced by another phenomenon that we also do not understand but that we can control. This notion of control is common to both the physical model that can be controlled mathematically and the homologue that can be controlled experimentally. It is through such control that we learn about the phenomenon that we hope to eventually understand. This approach has long been used in the design of such things as ships and aircraft, which could be scaled from the laboratory to the field, which means if we can determine how a scaled down replica of a vehicle responds in the laboratory to a turbulent fluid then we can anticipate how the full sized vehicle responds in the field to a proportionately larger turbulent field.

To many, engineering design is a less grand form of knowing than is pure science. Engineering has to do with building things and having them work in the way anticipated. It is this realm of knowledge that I will discuss in my brief remarks. I leave it to the other members of the panel and the audience to discuss the more

[1] Mathematics Division, U.S. Army Research Office, Research Triangle Park, NC, westb@aro.arl.army.mil.

elevated questions having to do with nonlinear systems theory, computability and stochastic processes.

Let us begin by making explicit what modeling in the hard sciences has meant traditionally. The paradigm of hard science begins with the strategy of Galileo and Newton in which physical phenomena are assumed to be continuous and differentiable. To these early scientists continuity and differentiability were not assumptions, but rather the result of observations of the physical world. The fact that material bodies must move continuously through space and time was self-evident. After all, how could a particle moving along a trajectory skip over a point between two adjacent points on a trajectory? It is also demonstrably true that experiments are reproducible and therefore the outcome of an experiment is only dependent on the initial state of the experiment, not on how that initial state is formed, that is, there is no memory of how the initial state is formed in the system dynamics - physical phenomena are Markovian. Therefore the goal of science is to construct analytical mathematical models of physical phenomena.

The analytical approach has been extremely successful in modeling the physical world. In the 19th century simple equations were used to model the propagation of sound and light, the diffusive transport of heat and particles and stochastic analysis using the mathematics of Sturm-Liouville. In this period the foundation of thermodynamics was also established using constitutive equations and equations of state. In the 20th century physical theory was generalized beyond the classical perspective of the macroscopic to describe microscopic phenomena using quantum mechanics. Each and every one of these models of physical reality could be mathematically characterized using the Sturm-Liouville theory of differential equations. The fundamental requirement for the application of this mathematical theory to physical phenomena is linearity. Therein lies the rub. Most physical phenomena rather than being linear are, in fact, nonlinear, at least in some domain of parameter values. Thus, all this analytical modeling of physical phenomena, as successful as it has been, is fundamentally flawed. Or more accurately stated, the domain of reality described by these models is much narrower than the scientists of these earlier generations believed.

One reason why such models have not been successful in the social and life (soft) sciences is that these phenomena are much more complex than those in the physical (hard) sciences and are therefore less amenable to traditional mathematical modeling. If inventors waited for complete understanding before developing such things as vaccines, modern man would have long ago been destroyed by plague and pestilence. The notion of predictability and control is common to all of science and through control we learn about the phenomenon that we hope to understand. So what makes a system complex? How is complexity of the soft sciences related to modeling and simulation of the hard sciences?

To answer these questions regarding complexity, let us review some history of technological achievements (Montroll 1987), using technology as a measure of one kind of knowing. The first achievement we point to is aviation. In 1903 the Wright brothers flew their plane at Kitty Hawk for a maximum time of 59 seconds. Ten years later Igor Sikorsky flew a plane with six passengers for six and one half hours. Twenty-five years after the first flight there was a regular passenger airline service that seated 38 passengers and which flew between Africa and India. The entire technology from prototype, to business as usual, took a quarter century. So if the measure of knowledge is the growth of an industry, we mastered aviation in 25 years.

Nuclear fission has had a similar success. The first fission chain reaction was successfully completed by Enrico Fermi in 1942 under the bleachers at the University of Chicago football field. It required only another fifteen years to complete the first commercial nuclear power plant. So nuclear technology is another field that we apparently mastered in short order.

Finally, there are the rockets of Goddard that successfully reached one mile high in 1933 and twenty-six years later there was the first Soviet cosmonaut orbiting the planet. We landed on the moon 10 years later indicating that the technology of rockets is certainly something that we understood.

Each of these examples represents a successful technological development in the last century and because we control the phenomenon, one would say that we understood it.

The last activity we mention is not an achievement, however, and that is nuclear fusion. The search for controlled nuclear fusion began with project Sherwood in 1951 and after over 50 years and hundreds of millions of dollars we still do not have fusion. Why? What makes fusion different from all the other technologies where we have succeeded?

More generally, what distinguishes the kind of problem that we were able to solve in the 19th and 20th centuries in the hard sciences, and the technologies we were able to develop, from those we were not able to solve? Is it the difference in the level of complexity of problems in the soft sciences from those in the hard sciences? Is complexity the reason that we have had such little success in solving the problems in the soft sciences? Saying that these unsolved problems are complex is in one sense avoiding the issue simply by giving it a name. I believe the reason for this failure is at least partly due to what the late Elliott Montroll called the tyranny of many dimensionless constants and we paraphrase a number of his arguments below (Montroll 1987). To understand the nature of this tyranny let us examine the description of fluid flow in three dimensions given by the Navier-Stokes equation:

$$\frac{\partial \mathbf{v}}{\partial t} + \mathbf{v} \cdot \nabla \mathbf{v} = -\nabla (p / \rho) + \nu \nabla^2 \mathbf{v} + \mathbf{F}/\rho. \tag{8.1}$$

Here $\mathbf{v}(r,t)$ is the fluid velocity at the space-time point (r,t), p is the pressure at the same point, ρ is the density of the fluid, ν is the viscosity (internal friction of the fluid) and $\mathbf{F}$ is an external body force such as gravity. We want to use Eq. (8.1) to introduce the notion of dimensionless constants, but do not worry, we have no intention of solving this equation.

To introduce dimensionless constants we take the NS-equation and transform it to an equivalent equation for dimensionless quantities. The velocity, pressure, and even gravity are all variables with dimensions - their numerical values depend on the units chosen. Suppose we wanted to investigate the flow pattern around a ship in the ocean. We would introduce a typical speed of the ship V, a typical length L

(say the length of the ship) and the pressure at the air-sea interface P. We can then divide out the dimensions using these quantities and obtain the time-independent equation in terms of dimensionless variables and three dimensionless constants:

$$\mathbf{u}\cdot\nabla'\mathbf{u} = -\frac{1}{E}\nabla'\mathrm{p} + \frac{1}{R}\nabla'^2\mathbf{u} + \frac{1}{F}\mathbf{e}. \tag{8.2}$$

Here $\mathbf{u}$ is the dimensionless velocity $\mathbf{u} = \mathbf{v}/V$ and the dimensionless constants are the Reynold's number R, the Euler number E and the Froude number F, given by

$$\begin{aligned} R &= V / \nu L, \\ E &= \rho V^2 / \Delta P, \\ F &= V^2 / Lg. \end{aligned} \tag{8.3}$$

In ship design, when typical values of the V and L are introduced into Eq. (8.3) both the Euler number E and the Reynold's number R are very much greater than the Froude number F and the body force term dominates in the dimensionless equation for fluid flow. Consequently, to first approximation, ship modeling can be based on Froude modeling; modeling with a dimensionless constant that depends on gravity. This scaling determines how to transform the results from a model in a ship basin to a full-size ship and establishes the power requirements to overcome the drag of the water on the ship. In airplane design the Euler or pressure term dominates, since it is the pressure difference between the bottom and top of the wing that determines the "lift" of the wing. A wind tunnel is the traditional device for measuring the lift and drag on a model airplane in a flow stream. Since the length L does not enter into the Euler number, the lift-to-drag ratio would be the same on a small airplane model as on a full-scale object of the same shape. This is what allows the design engineer to understand aerodynamics sufficiently well to construct the planes you all used to get here today. In each of the successful examples given only one dimensionless parameter dominated the phenomenon and consequently that parameter determined the corresponding technology.

So what about magnetically confined fission? Montroll (1987) argued that the magnetically confined fission programs have fallen victim to the tyranny of many dimensionless constants. The great engineering successes of the past have involved processes that could, to a first approximation, be characterized by a small number of dimensionless constants. Hence only a small number of model experiments were necessary to determine the feasibility of a project and to estimate the cost and difficulties to be surmounted. Even the space program was broken down into a number of subprojects, each of which could be analyzed in terms of a small number of dimensionless constants so that the results of many independent model tests could be used as a basis of the required full-scale engineering designs. The complication of the magnetically confined fission programs seems to be that all the hydrodynamic dimensionless constants (approximately eight) as well as several electromagnetic and nuclear dimensionless constants are intimately connected in the process of transforming a low density, low temperature plasma to a higher density, very high temperature plasma. There is only a very small region of this multidimensional parameter space in which one can successfully develop the desired technology - but where is it?

Are there any general conclusions we can draw from this example?

Let M be the number of dimensionless constants required to characterize a complex process. Then an experimental program must sample N_1 points along the axis of the first dimensionless constant, N_2 points along the axis of the second dimensionless constant, and so on. The total number of data points is then the product of the number of points along each of the axes in the M-dimensional space of characterization. The cost of the program, or the length of time it takes to reach understanding, is denoted by P and is proportional to the number of sampling tests:

$$P = kN_1 \times N_2 \times \ldots \times N_M . \tag{8.4}$$

Hence, if we define λ to be the average of the logarithm of the number of observations for each dimensionless constant,

$$\lambda = \frac{1}{M} \sum_{j=1}^{M} \log N_j \tag{8.5}$$

so the cost (time to understanding) increases exponentially with the number of dimensionless constants

$$P = k \exp[M\lambda]. \tag{8.6}$$

Thus, most complex phenomena require a long time and usually a great deal of money in order to be understood. A probabilistic argument similar to the one just given indicates that the probability of understanding a phenomenon, by going "directly to the point" in the development of a technology that involves M connected dimensionless constants decreases exponentially with the number of dimensionless constants. Thus, the number of dimensionless constants is a direct measure of the complexity of a phenomenon and therefore its knowability.

Certain social situations and environmental processes might also depend on a large number of dimensionless constants. It is timely to consider the training of soldiers and other military personal. This training or teaching may be characterized by a number of dimensionless constants, relating to everything from the biomechanics of physical training, to the cognitive abilities of learning, to the psychological factors for adapting to stress in the battlefield (West and Griffin 2003). The understanding of these processes is not exempt from the tyranny of many dimensionless constants; nor is an attempt to make policies exempt without a considerable insight into the manner in which a change in a single dimensionless constant influences others. Just as the enthusiast for magnetically confined nuclear fusion knows how he would like to solve the energy problem, so the enthusiast for social and environmental reform knows how he would like to make our lives full of harmony and beauty. Unfortunately, both of these classes of enthusiasts remain dreamers until the tyranny of many dimensionless constants is overcome.

Science has no special procedure by which it can overcome the tyranny of many dimensionless constants. The best it can do is to recognize a truly complex phenomenon where it exists and understand that it will require a long time and substantial resources to master this particular phenomenon. On the other hand, by identifying the number of dimensionless constants required to model or simulate a phenomenon, and how these dimensionless constants are inter-related, the scientist can distinguish between a truly complex phenomenon and one that is merely complicated. In this way the scientist can determine where to expect short-term returns on investments and where the waiting time will be substantially longer.

References

Montroll EW (1987) On the dynamics of evolution of some sociotechnological systems. Bull. Am. Math. Soc. 16(1):1-46

West BJ, Griffin L (2003) Biodynamics: Why the wirewalker doesn't fall. Wiley, New York

IIIB Organizational Issues of Uncertainty

9 A View from the Inside: The Task of Managing Uncertainty and Surprise

Erich K. Baier[1]

Let me shift the flow of the day a little bit, because I am not here to present a theory, or prove a theory. Actually, I come here with questions. Because I am not a social scientist; I am not even a natural scientist. I'm an engineer! As you know, engineers do things and reflect upon what happened afterwards. And this happened to me. I want to give you a little story of what we do so that you understand what gave me the insight to be here. Actually, it is still surprising to me that I am here.

The story goes back to the mid-nineties. As you might know, RISC architecture, which is the underlying architecture for UNIX, was actually invented and successfully brought to market first by IBM in the early nineties. Then IBM decided that this UNIX idea needed to take off; and the bigger the consortium in the industry, the greater our success would be. When IBM teamed up to do this, the team found itself as a distant last in the UNIX game in the middle of the nineties. At that point in time, DEC was the industry leader; Sun was growing; and Intel IA64 was threatening to take over the world. Also, at that point in time, there was a question. IBM took on a new CEO in 1993. We started that year with $16 billion in debt. And the questions were, "Well, when you are a distant last in the race, should you stay in the race? Should you buy out a competitor? (DEC was probably the most obvious one at that point in time.) Should you join forces? Or, should you just think about how to win the game?" And, actually, quite amazingly, what happened in this dire financial situation was that a group went forward to Lou Gerstner in 1996 and proposed to raise one and a half billion dollars to achieve high end UNIX leadership, not the next year, but by the end of 2001.

At that point in time, the response from the financial community in the corporation was, "Well, it doesn't make a lot of sense." And there were a lot of critics. If you look back, the program actually did happen in 2001, exactly to the month as predicted five years earlier. The most vicious competitor was bought out by Compaq, which has now bought out by HP - merged, I forgot to say. We also watched all the forces of that time, especially the dot com boom and bust, go by. I joined this group, the group that proposed to get IBM back into the UNIX game, in 1999. First the chip came out; then I came in as the guy who would tie the project all together to make a system out of it and get it to market. We now earn a good amount of revenue from that project. As amazing as it seems, we are there now. We launched this UNIX system in December of 2001, and since then, we have gone from a distant fifth to a very close second in the industry, according to International Data Corporation (IDC). We will announce our results next week. We might

[1] Director, UNIX Hardware Development, IBM Corporation, ebaier@us.ibm.com.

be doing better than second. And, as a result of this project, we have built up momentum and a team that has identity and pride. I cannot really describe it all, so I will be raising some questions based on my observations of this experience for us to discuss later.

So, let me start. The question here is, "How does a large organization deal with uncertainty, change, surprise, when in the corporate world, actually, the last thing you want to deal with is surprise?" This is the case because you are there to produce results; and you produce results at a predicted pace. And that brings me to the reason why I am here right now.

IBM has a good relationship with the academia, mostly with engineering programs through the IBM Austin Center for Advanced Studies. So we have grants through which we work with professors every year. And usually my colleagues go after engineering projects and engineering students. After what happened to me, as I described in my story, I began to think that there must be something, some science out there that deals with and tries to make sense out of what I had observed. And so I applied to support a research project on organizational complexity. First, they said, "Well, what's that?" And that brought me to David Gibson [from IC2 Institute] and then finally to the latest member of our tribe, Senem Güney [a Ph.D. student in the audience] who joined us more than a year ago. Since then, Senem has been trying to make us reflect upon her perceptions of what we are doing. And that helps us engineers a lot to start to think differently and pay more attention to what is actually happening around us. And maybe after this conference discussion, I will have learned even more. Then this will really be a very worthwhile exercise for me.

I have actually netted down complexity to four things I have observed. The first is the interaction of people (as groups). And by people in this context, I mean members of a team with distinct goals - like a marketing group, a sales group, a manufacturing group - and, how they interact to define a goal. Especially in our environment, due to the differences in focus, every group thinks within the frame of a different time span. Sales guys think about this quarter. Manufacturing guys think they need to make their monthly output. The marketing guy thinks, "How do I get a good story for the year?" And the developer thinks about what to do for the product in the next five years. So in this situation, as a developer, you are expected to establish a good dialogue and follow this great paradigm that says everything comes from market input. And yes, it comes. But it comes when it is too late. This is definitely not because of bad intent, but because of the organizational complexity coming from the varying focus of people's attention, you end up interacting on different wavelengths.

Second, we have the interaction of people with other people (as individuals). It's probably the interesting and nice thing that we don't know who will resonate with each other and who won't. An org chart does not tell you what is happening in an organization, or who to talk to in order to get to the gist of what is going on. What I observed once was quite interesting. After a certain while, you see types of characters that flock, and independent of where they are, they resonate and drive the thing. And then there is the second row of close followers that perceive themselves as leaders, but actually, if the first row of leaders goes away, they fall back.

They will not take over. This is something that I have observed very closely in my experience.

In the corporate life, efficiency and productivity are definitely variables that you need to keep an eye on. Also, whenever you try to do something new, you automatically create a perception that you are after people's status, or influence. In some cases, like when you see that efficiency can be achieved across different sites, soon you begin to be perceived as the guy who is the single, biggest threat to a particular site. This actually happened to me for a while, and I was probably "The Unwanted Number One" in another collaborating development site. This has gotten much better now that we are at a point where we can reflect. At the time, we were in this situation where it was almost impossible to sit back and have a dialogue about why we perceive certain things as threats. The problem was pushed away, as if it was always something that did not apply, or it was not the real reason. Now we can get to dealing with all that.

As I said before, in our environment, the market forces are constantly changing and new competitors are emerging. Especially in this project period that I mentioned earlier, where we saw the dot com boom blooming, a lot of people were questioning if we were on the right pace. But before we were done, the bubble burst. And even though we did not yet see the peak of the mountain, we are still here. And we are prospering quite nicely right now.

Also, we have strategic forces from inside, coming from people who think about corporate strategy, about what the corporate strategy should be and how it should pan out. This can create a lot of conflict. And when such conflict is visible on the level of groups who are goal-oriented and who want to drive the project, this can cause quite a bit of chaos. So, here's the question: "How do you adapt to these forces without turning into a reactionary chicken, which doesn't know which way to go, because each day shows a different direction?" In this case, you have to apply a certain sense filter, which is quite dangerous, because the sense filter can also block you from recognizing the small indicators that are so important. So, then the question becomes, "What is the right sense filter?" And I think, as we have learned this morning, what becomes important here is not the magnitude but the momentum of the change.

We also have budgetary forces. Even though corporations have deep pockets, we commit to a given budget for a program. The continued commitment to the budget is closely linked to the market success. This means that you might enter the year with half a billion dollars and might end up with only $400 million in your budget at the end of the year. So that is another kind of surprise you have to deal with. You can, if you want to, hire and fire people overnight, but that is something we don't want to do. We have learned another nice thing during the dot com boom. A lot of people said, "Well, let me find my fortune outside, and let's hope for a stock explosion somewhere else." And that actually gave us an opportunity to take on a lot of new students in the 2000 and 2001 time frame. Now I wouldn't want any one of those mercenaries who left, because I have built a really great and motivated team that has a sense of pride and accomplishment. When there is this sense, it doesn't matter if these people make less money than those who left.

Also, there is this nice thing about technology. The laws of physics apply to us all in the same way. This is a nice thing to remember when certain things sound a little bit strange. The laws of physics always apply, which is a great thing, and which makes us ask, "How do we deal with it right now?" in order to go through a given program. We build a prototype concept, which is a proposal to pursue a project. Then we apply certain development criteria, which are supposed to match our task format - we are in the computer business, not the airplane business, and we don't want to turn suddenly into an airplane builder or something like that. Then we launch this proposal and develop the concept. Then we see if the concept has market potential, if it maintains competitiveness, and all that. You can have a margin of error of about twenty percent at that point in time. Then, we say, "Well, does this make sense?" and we develop a plan. At this point, we are getting serious. This means you have done the plan to get to roughly five percent accuracy and you commit yourself to the revenue plan of the corporation. From that point on, the "no surprises" part starts. And from that point on, my job is to go to the senior executives in the northeast every month and tell them that nothing has changed and that we are still on track. And when we are done, we say, "Did we end up with what we thought we would build, and is this the thing we want to launch in the marketplace?" It is possible to decide at that point in time that the launch shouldn't happen. This has happened once, and we actually stopped a program. By the end of the life of the project, or in the after-life, we reflect back to see if the assumptions have been materialized; what forces caused the assumptions to change; and what we have learned out of it.

Consequently, one of the major challenges we have in our business is the balance between the stability, predictability and adaptability to change. So, one of the questions I want to throw out to the room here is, "How are these linked?" As you can see, surprise is usually a deep-structural event in this environment. However, in the corporate world, surprise is something that is almost unwarranted. The response to surprise is almost an accusation: "Well, if you are surprised, you didn't do your homework." The corporate culture usually resists change. And that's where, I think, the danger is for a lot of corporations who push back on signals to change and then run over the cliff.

So, to sum up my observations, I will say that complexity in organizations is mostly shaped by the following forces: The interaction among people in their group identity, the interaction among people as individuals, external forces and (internal) strategic forces as well as the way the corporate climate allows you to deal with these forces. Then my last question is, "How can we bring a better understanding of these forces into a corporation so that, outside the social sciences, we can understand and deal with their interdependencies a little better?" Even as engineers. Thank you.

10 An Introduction to the Mathematics and Meaning of Chaos

Larry S. Liebovitch[1]

It is a pleasure for me to be here, both to talk and to listen. I thank Curt Lindberg and the other organizers for this opportunity. It has been a long time since I have been here in Austin. In fact, the last time I came here, from Boston; I got here by flying on Eastern Airlines. So it's been some time.

What I would like to do is to try to knit together some of the more mathematical approaches we have heard this morning; and the approaches that we are going to hear later. The question that I want to ask is: What is the nature of the relationship between the mathematics and the metaphorical lessons that we learn from it?

I'll start off by giving some very simple mathematical examples for people who are not familiar with the mathematics. Here I present two different data sets; two different time series. You can think of these, for example, as temperature measurements made each day over a two-month period. Although these data sets are different, they both seem to have the same qualitative feature. They both look as if the temperature each day varies at random.

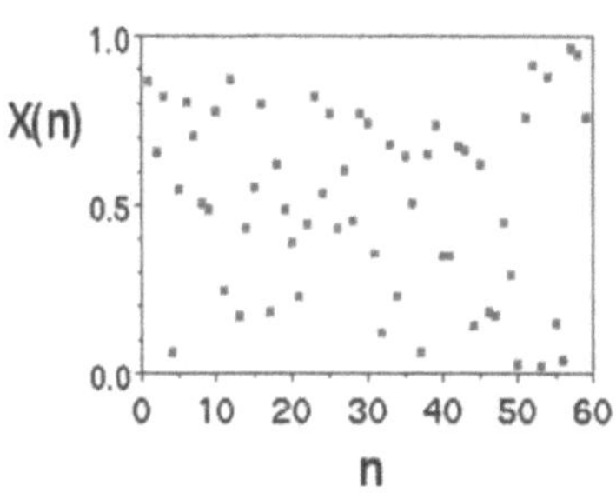

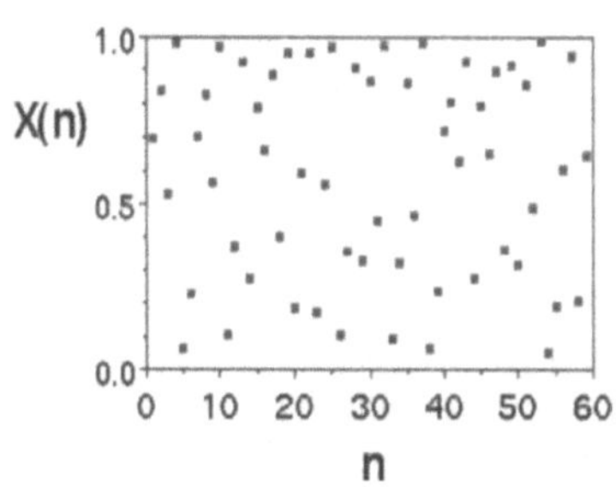

But what we are learning today is that not everything that looks random is random. You can see this when I transform this data. Instead of plotting the temperature each day I can make a plot of temperature today against the temperature yesterday. If I do that for both these data sets, they look something like this.

[1] Professor, Center for Complex Systems and Brain Sciences, Center for Molecular Biology and Biotechnology, Department of Psychology, Florida Atlantic University, liebovitch@clifford.ccs.fau.edu.

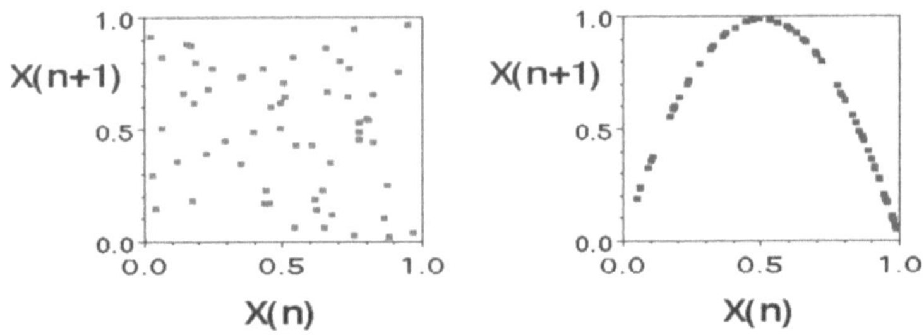

This first data set really does look random. The temperature today has no particular relationship to the temperature yesterday. It varies all over the place. But if I look at the second data set, we suddenly see that it is not random at all. The temperature today is a very simple function, expressed by this line, of what the temperature was yesterday. Even though we understand how the temperature, one day, affects the temperature the next day, over many days the temperature seems to fluctuate at random. The jargon word for describing this phenomenon is the word "chaos."

We have learned something new and important about the data by transforming the time series of the temperature every day, to this plot of the temperature today versus the temperature yesterday. This plot of the temperature today versus the temperature yesterday is called a phase space.

Let me give you another example. This is one from Lorenz, and it's a model of the motion of the air in the atmosphere. In this model, the air is heated from below; and then cooled from the top. The hot air rises and the cool air falls. The motion of the air forms convective rolls that rotate. The first one on the left starts off rotating counter-clockwise; and it will rotate that way for a while. It slows down and speeds up. It is actually going to stop. When it stops rotating, it may rotate in the opposition direction. It does that for a little while. Then it stops; slows down; and speeds up; and stops. And then it may switch again and rotate in the opposition direction.

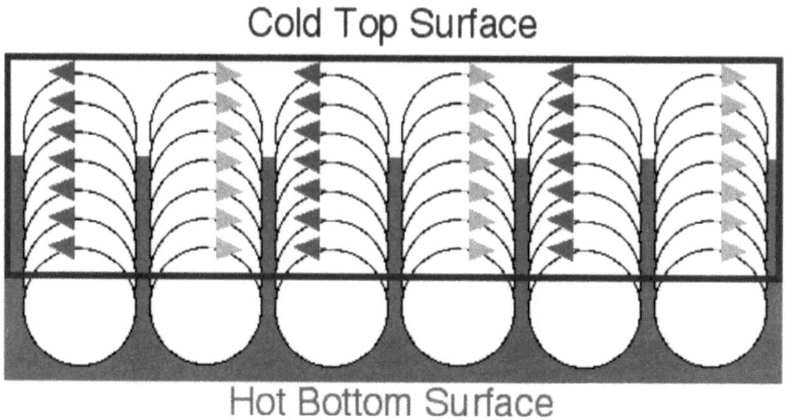

We can make a plot of X representing whether the rotation is clockwise or counter-clockwise. When X is positive the rotation is clockwise, when X is negative the rotation is counter-clockwise. Let's say that we start off with the value of X equal to 1 so that the air starts off rotating weakly clockwise. Then it will switch to rotating counter-clockwise; rotating clockwise; and then counter-clockwise, and so on. And it keeps switching from one rotational direction to another. These switches seem to happen at random. In one case, I started off the system with the value of the amount of rotation exactly equal to 1.00000. Now, let's say I do it all over again, but this time I start off in almost exactly the same way; with the initial X not exactly 1.00000 but the initial X equal to 1.00001. Pretty close. And as you might expect, the second time the air starts off doing the same thing as it did the first time. But after awhile, the second system is doing something very different than the first system did. In fact, at one point, when the first system is rotating clockwise, the second system is rotating counter-clockwise. They are doing opposite things!

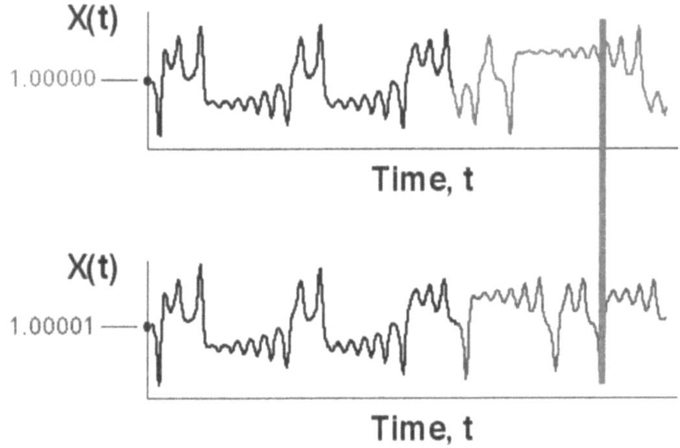

This is something that we are not really used to seeing. If we have a team of people we know well and we give them a series of instructions on Monday, and they carry out their task in a certain way, we really believe that if we gave the same people the same series of instructions on Wednesday, they would carry out their task in pretty much the same way. That is not true for the Lorenz System. And it's not true for chaotic systems. We can have the same system, with the same mechanics or the same people, and the same instructions; but get wildly different results.

The jargon words for this are: "sensitivity to initial conditions" which is also called the butterfly effect. If the butterfly beats its wings in China, it will change the weather five days later in New York. Everyone seems to have their favorite place where the butterfly beats its wings and where the weather changes. A small sample, from a few articles and books that I looked at include: China and New York, Peking and New York, Beijing and the Atlantic, China and the Caribbean, Java and Chicago, (anywhere) and Indonesia, Brazil and Texas, Brazil and Alaska, California and Nebraska, Gliwice Poland and Boca Raton Florida (this is from my book, Fractals and Chaos Simplified for the Life Sciences, 1998, Oxford University Press, I have friends in Gliwice and live in Boca Raton).

Here are two more examples of the amount of rotation of the air versus time for two different computer simulations of the Lorenz System. Both simulations come out different. Again, this is pretty surprising to most people who are used to handling data. We are used to thinking that, if we run an experiment, we should pretty much get the same result every time that we run the same experiment. Yet here, every time we run the same experiment, we get different results.

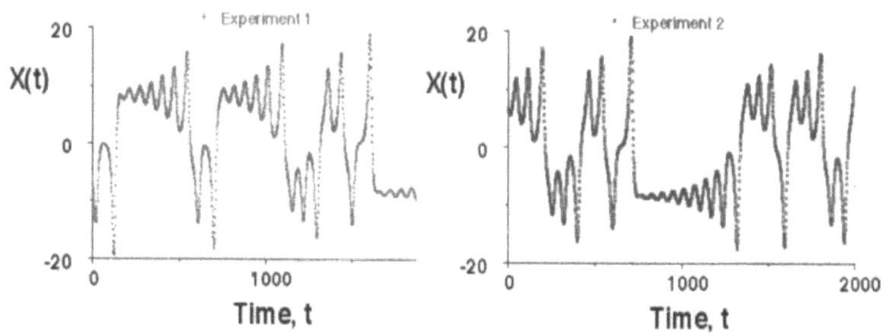

In a certain sense, the time series here is not the most real thing about the experiment. We can see that by doing the same transformation here to create a phase space, as we did before in the temperature measurements. This is what that phase space looks like. It has an object in it, which is technically called a strange attractor. And what you can now see is that each of these two experiments forms just slightly different and sometimes even overlapping parts of that same object. So even though the time series is different for each experiment, the attractor is the same every time the experiment is run.

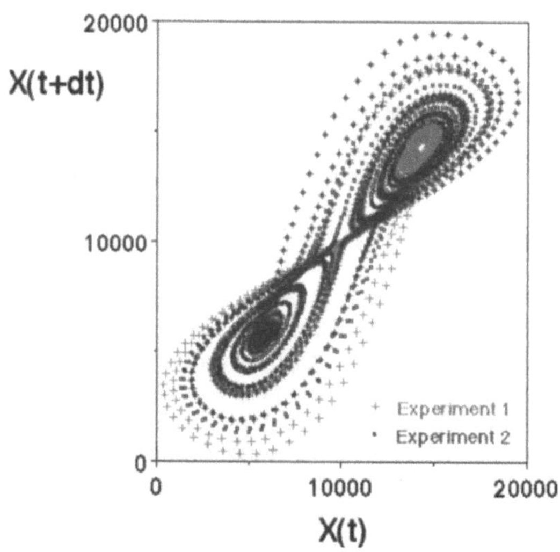

In a certain sense, what is really real here is the attractor; not the time series. The main points here are: 1) What is real is not the values as they progress in time, but a transformation of those values into the phase space, 2) The sensitivity to the initial conditions, which means that even if we understand the details of what is happening the outcome can still be uncertain. These new ideas have changed how we think about things in the world. What we are hearing at this meeting is what I would call metaphorical applications of these ideas. The mathematics, although I haven't shown it to you here, is rigorous and concrete. The issue here is Can we apply these mathematical ideas without the rigor of the mathematics?

The metaphorical lessons that come out of chaos are that you should stay loose and plan as you go along; you should be prepared to adapt to changes because the sensitivity to initial conditions means that there will always be unforeseeable consequences, cause and effect will not be so easy to determine, and to be successful you need to stay nimble and dynamic.

I think this is a "good news; bad news" story. So let me tell you the "bad news" first. The bad news is this is not a direct application of these real mathematical ideas. But, if you just applauded, there is actually more good news than bad news with this. I think this is a valid application of the mathematical concepts in a metaphorical sense. Some of the problems that we deal with are very, very difficult and so anything that gives us new ideas to help us understand them is worthwhile. At the very least, these ideas give us justification for doing what we want to do anyway; what we know is right, and what we know will work.

This brings us to the issue, How do we know it will actually work? I deal with some issues in Education; one of which is "objective assessment." This reminds me sometimes of a poem, 'The Circus Animals Desertion' by William Butler Yates where he writes: "Players and painted stage took all my love, and not those things that they were emblems of." Do we sometimes mistake emblems for the things of which they are emblems of? In Education, do we mistake test scores and grades for knowledge and understanding? What's very popular now in Education are these assessment questionnaires to assess both faculty and students. This has always struck me as being odd. If you have a problem in a relationship with your kids, or your boss, or your lover, you don't give them a multiple-choice questionnaire to fill out. You deal with them through some human interaction. And maybe that's the most important thing that goes on in Education. Maybe in assessing things in Education, we have mistaken emblems for things of which they are emblems of.

This brings us to a question that someone raised from the floor, which is "What is the criteria by which we judge whether we have been successful?" If we have one company that has low profits; happy, creative, productive employees; and makes a useful product that many other people use; is this company more or less productive than another company that runs very high profits for eight consecutive quarters and then goes belly up? In the world in which we live, there are a lot of criteria that would judge that second company as being more successful, at least for the first eight quarters.

So the issues that I would like to raise are: First, Is it fair to apply these mathematical ideas in a metaphorical way? And second, How do we tell if these ideas have been successful?

Let me tell you where you can learn more about these ideas. I am embarrassed that I am the first person to plug their own book here. But I am going to do it anyway. I did a book a few years ago called *Fractals and Chaos Simplified* for the *Life Sciences* (1998, Oxford University Press). What is unusual about this book is that it consists of facing pages. The left one is always text, and the right one is always pictures. It leads you one thought at a time through these ideas in a way a qualitative way with words but, as one reviewer said, it is still "mathematically honest." And the other thing we have been developing, funded partially through the National Science Foundation, is a set of curricula materials that use fractals and chaos to illustrate what mathematics is, for a college mathematics course for students who never liked and never did well in mathematics. That's nearing completion now and if anyone is interested using those materials to teach such a course, we can send you an "almost finished" free demo (http://www.ccs.fau.edu/~liebovitch/larry.html). Thank you.

IIIC Fundamental Uncertainty and the Delivery of Health Care

11 The Social Construction of Uncertainty in Healthcare Delivery

James W. Begun[1] and Amer A. Kaissi[2]

We explore the following question: How would healthcare delivery be different if uncertainty were widely recognized, accurately diagnosed, and appropriately managed? Unlike most studies of uncertainty, we examine uncertainty at more than one level of analysis, considering uncertainty that arises at the patient-clinician interaction level and at the organizational level of healthcare delivery. We consider the effects of history, as the forces and systems that currently shape and manage uncertainty have emerged over a long time period. The purpose of this broad and speculative "thought exercise" is to generate greater sensemaking of the current state of healthcare delivery, particularly in the realm of organizational and public policy, and to generate new research questions about healthcare delivery. The discussion is largely based on experience in the United States, which may limit its generalizability.

11.1 Defining Uncertainty

Uncertainty means doubt, a lack of sureness. In quantum mechanics, uncertainty refers to lack of sureness about geometrical position and speed of moving particles. Heisenberg's uncertainty principle states that the more precisely we determine the position of an electron, the more imprecise is the determination of velocity in this instant, and vice versa. In the functioning of systems, uncertainty can be contrasted to determinism. In deterministic systems, functioning of the system is sure or certain, because evolution of the system is governed by a set of rules that, from any particular initial state, can generate one and only one sequence of future states (Prigogine 1997:201).

Uncertainty derives from several sources, most notably complexity and lack of knowledge. Complexity produces uncertainty because precise prediction in complex systems is impossible. Nonlinear interactions among system components mean that bifurcation and choice exist within the system, leading to the possibility of multiple futures and creative or surprising responses (Allen 2000:79), in contrast to the deterministic system described above. Consistent with this notion, in reference to healthcare organizations, uncertainty has been defined as the inability of an organization to accurately predict the consequences of an action or the future

[1] James A. Hamilton Term Professor, Department of Healthcare Management, Carlson School of Management, University of Minnesota, jbegun@csom.umn.edu.

[2] Assistant Professor, Department of Health Care Administration, Trinity University, amer.kaissi@trinity.edu.

state of an organization and its environment (Shortell and Kaluzny 2000:468). Generalizing this definition, we define uncertainty as "*the inability of agents in systems to accurately predict the consequences of an action or the future state of the agent, the system, or the environment.*" Uncertainty is a perceptual or subjective phenomenon in part because it is processed by human beings and aggregates of humans (e.g., organizations or legislatures). There is a boundary of precision that humans can never cross (Clampitt and DeKoch 2001:54). This results in "objective" uncertainty being changed in ways that are difficult to foresee and to understand. We emphasize these psychological and social uses of perceptions of uncertainty throughout this paper.

Healthcare delivery commonly is characterized as "highly uncertain." Such characterizations vary in meaning, and they span a range of levels of analysis and types of uncertainty.

11.2 Types of Uncertainty

As noted above, some component of uncertainty arises from lack of knowledge - knowledge that reasonably can be expected to reduce that uncertainty. An example is uncertainty over the diagnosis of an acute disease with a specific and identifiable cause. A throat culture may validate the diagnosis of strep throat, for instance. However, much of uncertainty is fundamental - it results from the nature of complex systems, and will not be reduced by an improved information base and improved management of information. An example would be uncertainty about the future paths of an illness, particularly a chronic illness, due to the complexity of chronic illness processes. As number and diversity of elements in an interdependent system increases, the chances for surprise increase. At the organizational level, complex arrangements among subunits or organizations and external agents to deliver healthcare ensure the emergence of surprise and therefore fundamental uncertainties.

Whether fundamental uncertainty is much more common than uncertainty that can be reduced through information, as some argue, is an interesting question. Much of uncertainty falls in a middle category, in that uncertainty can be reduced but not totally eliminated through more information. For our purposes, the distinction between fundamental, irreducible uncertainty and reducible uncertainty is taken to be a useful one.

11.3 Levels of Analysis

Uncertainty can be analyzed at different units of analysis, exploring uncertainty as perceived or defined, e.g., by the individual, dyads, groups, organizations, clusters of organizations, communities, and societies. At the individual level, individuals face uncertainty about the extent to which they will suffer illness, and individual clinicians face uncertainty about the extent to which they can be of assistance. At the micro level of healthcare delivery, patients and clinicians carry their assumptions about uncertainty to the patient-clinician interaction. Perceptions of uncertainty that emerge in the interaction, and therefore characterize the interaction, become managed by new social processes. Healthcare delivery workgroups and organizations face uncertainties related to their ability to coordinate complex arrangements to produce high quality and seamless care. Individual perceptions about uncertainty come to be reflected in cultural values within organizations and societies, and within organizational policies and strategies, and public policies. In turn, these social institutions affect the way that individuals perceive uncertainty. As macrosystem dimensions, the ways that societies manage uncertainty through cultural values and policies are a critical lever for change throughout the lower level systems.

11.4 Uncertainty as Socially Constructed

In this discussion, we emphasize that perceptions of uncertainty are shaped by social, economic and political forces. We refer to this as "the social construction of uncertainty." At the individual level, observers are influenced by their own positions, education, and economic and ideological interests. At the interaction level, control over defining and reducing uncertainty in the interaction is a source of power and reward. At the organizational level, specifying and treating uncertainty becomes a management tool, reflecting pressures on the organization and its managers. The social construction of uncertainty leads to questions such as, Who owns the uncertainty definition enterprise? And who manages uncertainty reduction? What are their interests? How were they trained and socialized?

We address these questions in the case of individual interactions between patients and clinicians, and in the case of organizations delivering healthcare to individuals. We probe variations in 1) perceptions of the level of uncertainty, 2) perceptions of the degree to which uncertainty is fundamental vs. reducible, and 3) strategies and policies for managing uncertainty.

11.5 Managing Uncertainty at the Patient-Clinician Level

Most consumers of healthcare services expect and accept a certain degree of uncertainty in the emergence and patterning of their illnesses, including both the appearance and course of illness. Perceived uncertainty is both fundamental (we don't know how our illness will progress or when another one will emerge) and knowledge-based (sometimes we can identify and appropriately treat illness through greater knowledge of causes and interventions). Individual patients bring such perceptions of uncertainty to their interactions with clinicians.

In managing uncertainty in the interaction between clinician and patient, the assumption typically has been that the patient is ignorant, and the clinician has the answers. Informational asymmetry characterizes the physician-patient relationship with the assumption that the physician is the holder of the information needed to reduce uncertainty, which the patient cannot know or understand. Medicine attracts and reinforces individuals who believe in science and the power of knowledge to overcome disease. Typical clinician-patient conversations are heavily dominated by the direction of the clinician (Mishler 1997; Waitzkin 1989). In addition, the clinician generally is expected to minimize the degree of fundamental uncertainty ("We will get to the root of this"), though in part this is rationalized to be therapeutic (because it is reassuring to the patient).

The management of uncertainty by clinicians is supported at successively higher levels of systems. Healthcare delivery organizations typically have been structured as "physician workshops," where the physician, not the patient, is the main customer. The healthcare system has evolved a rigid structure of professions that are expected to self-regulate, develop specialized knowledge, and apply that knowledge for the benefit of consumers. At the center of the system of professions is the physician, who was accorded professional status through licensing laws in the early 1900's, aided by the Flexner Report and subsequent raising of standards in medical education. The scope of medicine created a sweeping monopoly over most of healthcare delivery at the clinical level through its linkage to science and through effective strategizing and political action (Starr 1982). This was a defining event in the evolution of the management of uncertainty. Other clinical providers that have emerged in response to new needs and demands, new knowledge and technology, build their activities on a terrain that structurally is very rigid and controlled (Begun and Lippincott 1993). In general, less prestigious and less lucrative task domains have been carved out around the profession of medicine.

While individual clinicians have accepted a role as "reducers of uncertainty," most at the same time recognize and accept fundamental uncertainty as well, as reflected in this passage:

> There can be uncertainty about whether an individual is well (self-limiting episode) or ill (serious risk of medical impairment of health without medical intervention). If ill, there can be uncertainty about what is the correct diagnosis. After the correct diagnosis is determined, there is often uncertainty about what is the most appropriate treatment. Even at this point, there may be still uncertainty about the effects of treatment. This relates not only to the possibility of an untoward reaction to the therapy but also to the uncertainty in predicting the degree of recovery from illness, even with the correct diagnosis and the appropriate treatment. Thus, decision making under uncertainty is inherent to patient management and minimizing uncertainty is an integral part of the practice of medicine. (Wingert 1996)

Physicians D. M. Mirvis and C. F. Chang (1997:385) echo this perspective: "Uncertainty is a fact of life in medical practice. We, as physicians, usually do not know unequivocally the full extent of a patient's disease and even less often do we know the best single approach to diagnosis and treatment, if one were to exist even in theory." While much of the uncertainty is attributed to lack of knowledge about disease and therapies, some is due to biological (and other) variability, meaning that "it is unreasonable to expect that a single path for diagnosis or treatment exists for all patients. That is, uncertainty will remain" (Mirvis and Chang 1997:385).

Professionals walk a fine line between arguing for the existence of uncertainty and providing service to reduce it, because their legitimacy depends on societal acceptance of a degree of uncertainty in a work domain, but not so much that specialized knowledge cannot help. Professions need to create a domain in which their work is somewhat uncertain, or else professional work can be done by technicians. Acceptance and manipulation of a degree of indeterminancy has played an important role in the development of the profession of medicine and other similar professions (Jamous and Pelloille 1970).

As a result, professionals are motivated to both reduce uncertainty through the application of expertise, and maintain uncertainty that justifies their monopoly of a work domain. In this sense, professions are a primary institution through which society both recognizes uncertainty and pursues its reduction through information. The centrality and power of the physician role has widespread consequences for how uncertainty is handled at levels above the patient-clinician interaction, in organizations and public policy. In addition to control of information flow (vs. greater consumer input and access), it has contributed to physician-dominated (vs. team) treatment, and societal investment in "physician-friendly" research and medical products. A huge investment in medical research is sponsored by governmental and quasi-governmental agencies in the United States, most notably the National Institutes of Health (NIH). A clear goal of the NIH is to improve health through the development of new knowledge. Only recently has attention turned to such issues as the *dissemination* of knowledge, which is a complex social process, or to developing the knowledge base of non-traditional providers and providers other than physicians (e.g., the National Institute for Nursing Research, and funding of complementary therapy studies, began only in the past decade). Attention has yet to turn in a serious manner to developing knowledge on coping with disease and mortality (fundamental uncertainties) and on managing the delivery process (uncertainties caused by complexity of organizational arrangements). At issue is the balance of societal commitment to the reduction of uncertainty about medical disease, vs. investment in other forms of new knowledge. This has to do with political control of agenda setting, which traditionally has devolved to medicine, rather than failure to "see" uncertainty. The issue is who gets to define how that uncertainty is managed.

In a society where patients and clinicians are optimistic about reducing uncertainty, a healthcare product that promises acute relief or a cure is more likely to sell. Pharmaceuticals and medical devices meet this demand, and have spawned a powerful set of businesses that then reinforce this approach to reducing uncertainty in healthcare delivery.

What alternatives might emerge if fundamental uncertainty were more widely accepted in the patient-clinician interaction, and if the delivery system were not constrained by a rigid professional hierarchy? At the clinical level, uncertainty management is a joint responsibility of clinician and patient, and should be treated as two-sided. The patient knows more about his or her illness than typically is communicated (e.g., about their medical history, related bad habits, side effects). The emerging model of clinician-patient relationship, particularly with chronic disease, emphasizes patient self-management and co-management, and extensive collaboration between patient and physician. In this context, listening is a key and underdeveloped physician competency (Mechanic 1996). A partnership framework promises to maximize the acceptance of fundamental uncertainty and to more effectively reduce informational uncertainty.

Table 11.1 summarizes our discussion of differences between traditional strategies and policies managing uncertainty, which focus on uncertainty reduction, and strategies and policies that would embrace and accept uncertainty.

Table 11.1. Strategies and policies for uncertainty management at the patient-clinician level

To reduce uncertainty	To embrace uncertainty
Attention to cure	Attention to care, prevention, coping
Clinician as objective, detached	Clinician as subjective, connected
Scientific medicine	+ Complementary medicine, spiritual medicine
False hope	Honesty
Clinician controls information flow	Balanced information flow

11.6 Managing Uncertainty at the Organizational Level

We turn to the second key level of uncertainty management in healthcare, the organizational level. Here we examine the uncertainties that arise within organizations that deliver healthcare.

There is no doubt that healthcare delivery by organizations is extraordinarily complex, if not uncertain. The work of over 300 different professions has to be coordinated and managed, under the lens of extensive external regulation and community oversight. This complexity has been cogently summarized as deriving from "curtains" between four different cultures of care, cure, community, and control (Glouberman and Mintzberg 2001a, 2001b).

Healthcare managers typically attack this source of uncertainty with traditional management solutions (Begun, Zimmerman and Dooley 2003; McDaniel 1997). To the extent that the problems result from extreme complexity and are fundamental rather than reducible, the traditional tools may aggravate rather than ameliorate uncertainty. At the organizational level, uncertainty identification and reduction is typically defined as good management. Traditionally, leaders uncover and "fix" uncertainty (Stacey 1992). Illustrating the subjective nature of uncertainty in organizations, leaders are taught to engineer perceptions of uncertainty in order to create the foundation for organizational change (Kotter 1995:60). General Electric was transformed when Jack Welch found too many of its product lines comfortable and complacent. He engineered a "sudden confrontation with the inexorable hazards of natural selection" by requiring that every business be number one or two in its industry (Pascale, Millemann and Gioja 2000:27). Similarly, healthcare executives in the 1980's were warned that capitation would soon dominate the reimbursement landscape. The message was that managed care was coming, and uncertainty could be reduced by vertically integrating to manage the whole patient. Healthcare leaders, perceiving (unrealistically) high environmental uncertainty (e.g., the specter of a fully capitated payment environment), commenced and built the wave of mergers, acquisitions, and integration in the 1980's and 1990's. In hindsight, this was partly an overreaction, as many health systems are disintegrating (Friedman and Goes 2001; Luke and Begun 2001). The tendency to mimic the strategies of other organizations in order to reduce uncertainty may be stronger among healthcare organizations than organizations in many other settings (Bonabeau 2003; Fiol and O'Connor 2003).

As with physicians at the consumer-clinician interaction level, professional managers in healthcare delivery organizations wield significant power in designating levels of uncertainty and how they are managed. Management in healthcare delivery has been professionalized since at the mid-1900's, shared in part with clinicians (largely physicians), and substantial pressures continue to further professionalize the field of healthcare administration (Griffith, Loebs and Dalston 2001). Seminal events in the management of uncertainty by healthcare managers include the formation of professional associations (American College of Healthcare Executives in 1933, Association of University Programs in Health Administration in 1948, Accrediting Commission on Education in Health Services Administration in 1968), and the continued growth of healthcare management education programs.

We have argued elsewhere (Begun and Kaissi 2004) that organizational leaders in healthcare shape their perceptions of uncertainty to rationalize the suboptimal performance of healthcare delivery organizations and to exaggerate the challenges of managing healthcare delivery. Many potential explanations can be provided. High uncertainty may be a rationalization promoted by some healthcare managers to justify the performance of healthcare organizations. Healthcare organizations are subject to criticism for high costs, limited access, and less-than-stellar safety and quality records. A natural defense of the criticism may be for managers to stress the difficulty of managing healthcare organizations, a difficulty that is amplified by perceived uncertainties about regulation, reimbursement, technology, and other environmental elements. A second explanation is that uncertainty and uncertainty reduction are historical traditions or myths passed on by generations of managers. This is supported by the fact that virtually all eras of healthcare since the 1950's have been described by practitioners as "extremely volatile" and "unprecedented" (Begun and Kaissi 2004).

This is similar to the argument that clinicians use to justify self-regulation. That is, it's good business to see uncertainty, then reduce it. It's tougher business, or bad business, to see uncertainty and coexist with it or be slowed by its complexity.

How would healthcare delivery be different if uncertainty were recognized and expected, rather than oversimplified and avoided? Table 11.2 summarizes ideas that observers have presented to improve managers' abilities to accurately "see" uncertainty and to more effectively embrace it. Generally, organizations that embrace uncertainty would engage in more experimentation and learning, have loose connections in addition to tight ones, emphasize culture and participation as control mechanisms rather than formalized and centralized structure, and have top managers whose role is to "make sense" rather than "make decisions."

Table 11.2. Strategies and policies for uncertainty management at the organizational level

To reduce uncertainty	To embrace uncertainty
Simplify, standardize, formalize	Complexify, experiment, learn
Emphasis on structure	Emphasis on culture
Mistakes made by individuals	Mistakes made by systems
Tight integration	Loose-tight integration, e-merger
Manager as decisionmaker	Manager as sensemaker
Control information flow, participation	Increase information flow, participation

Boisot and Child (1999) and Ashmos, Duchon and McDaniel (2000) describe many of these traits as those of complexity absorption, as opposed to complexity reduction. Complexity reducing organizations seek order, in the same way that they seek certainty. Complexity absorbing organizations facilitate information exchange and allow generation of multiple interpretations of information. Information flow and exchange is facilitated by widespread participation in the organization's decision-making (Ashmos, Duchon, McDaniel and Huonker 2002).

Another conceptualization of organizations of this type labels them "mindful" organizations (Weick and Sutcliffe 2001). Mindful organizations are prepared internally to identify and deal with surprise. Mindful organizations analyze their failures in order to learn. They are in the habit of looking for complex rather than simple interpretations of events. Mindful organizations are extra-sensitive to operations, so that perturbations - anything out of the ordinary - can be noticed as early as possible. An organization's commitment to and experience with resilience reminds members that the organization will face surprise, will on occasion fail, but will learn and ultimately will survive. It also gives members "permission" to be innovative when faced with unpredictable adversity.

Applied to healthcare organizations that work together, such a management style has been described as representing "emerger" rather than merger (Zimmerman and Dooley 2001). Synergy naturally evolves from joint activities, rather than being pre-planned and imposed. Relationships and connections are managed, and syngergies emerge through self-organization.

Managers in organizations that accept fundamental uncertainty assume new roles and require new skills (McDaniel and Driebe 2001). Of these, sensemaking is a critical one. Recalling our definition of uncertainty as "the inability of agents in systems to accurately predict the consequences of an action or the future state of the agent, the system, or the environment," it is clear that many management activities in healthcare are uncertain, and often are irreducibly uncertain. In that context, meaning comes not from knowing what is going on, but from making sense of it. This subtle but powerful realization transforms the role of the manager.

11.7 Comparison of the Two Levels of Analysis

A comparison across the two levels of analysis of uncertainty management in healthcare delivery reveals several similarities. First, there is significant consensus at both levels that much of healthcare delivery is uncertain. Clinicians, patients, and managers of healthcare organizations are aware that much of healthcare delivery is unknowable and unpredictable. However, the gap between seeing this uncertainty and appropriately managing it is large. Second, professionals are in a critical role as arbiters of uncertainty definition, identification, and reduction. Medical professionals and professional managers maintain significant control over the social construction of uncertainty. Third, uncertainty levels and uncertainty management mechanisms emerge and change in response to the social control interests of professionals. Discovering and then reducing uncertainty is "good for business" in medicine and in healthcare management. The ongoing cycle of uncertainty "discovery" and "reduction," then discovery of new uncertainties, maintains power of the professions. This places control of uncertainty diagnosis and management as key nodes in the uncertainty management system.

The patient-clinician interaction level and organizational level differ in significant ways, as well. The profession of medicine has a longer and stronger monopoly on the process of uncertainty reduction than professional managers do. Professional managers in healthcare delivery organizations share responsibility and power with a host of external regulators, clinicians, and board members. The profession of medicine is less interdependent with other disciplines in the management of uncertainty. This is changing, much to the chagrin of many physicians, as delivery organizations and payers impose more intrusive monitoring and control over care processes.

11.8 Strategies for Change in the Management of Uncertainty

To achieve change in the management of uncertainty at the patient-clinician level, a special onus is placed on medical professionals, as they can be significant barriers to change, or creators of transformation. As science accepts fundamental uncertainty, physicians who lead in interpreting and implementing the practical implications of this can have a powerful impact on their profession and the system of healthcare delivery. Changing the culture of physician education that targets minimizing uncertainty is a critical lever for change.

At the organizational level, a key to change is the minds of organizational "leaders." They have a key position in the system of uncertainty interpretation and reduction, and can be barriers to change or facilitators of transformation. Management education needs to include courses that expose students to the dangers of oversimplification of complexity, and overapplication of traditional management techniques to complex problems.

As new professions emerge in healthcare delivery, and as existing occupations seek status as professionals, they are in position to more accurately perceive and treat uncertainty. They are less subject to the constraints of tradition. As Wilsford (1994:251) explains, most actors in the healthcare system "are hemmed in by existing institutions and structures that channel them along established...paths." New leaders and new professions promise a way out of the traps of the past.

11.9 Strategies for Research on Uncertainty

Table 11.3 lists a number of research questions that derive from an interest in and understanding of uncertainty. First, empirical descriptions of perceptions of uncertainty are rare at both the patient-clinician level and the organizational level. (A recent exception is Schor, Pilpel and Benbassat [2000].) The perceptions of fundamental uncertainty that patients, nurses, physicians, and other clinicians bring to their interactions need study, as well as those of managers. The extent to which these perceptions are shaped by educational programs is also important. Of particular interest is the difference between nurse and physician education and socialization, as nurses may well be better prepared for the complexity of healthcare delivery than physicians are. In addition, variations in perceptions of uncertainty by profession and management setting can be expected.

Table 11.3. Research topics on fundamental uncertainty in healthcare delivery

Levels of perceived fundamental uncertainty
Patients
Clinicians (e.g., physicians and nurses)
Managers
Factors affecting levels of perceived fundamental uncertainty
Content of educational curricula and socialization experiences
Work setting (e.g., hospital vs. clinic)
Consequences of variation in levels of perceived fundamental uncertainty
Effectiveness of patient-clinician interaction
Processing of medical errors
Patient satisfaction with interaction
Effectiveness of organization
Processing of management mistakes
Levels of participation
Levels of information flow
Learning and experimentation
Role of manager as sensemaker
Satisfaction of clinician, manager
Quality of worklife
Job tenure
Career progression

Second, the consequences of varying perceptions of uncertainty need to be analyzed. We have implied that more accurate and realistic acceptance of fundamental uncertainty will be reflected, for example, in greater information flow between clinician and patient, more honest discussion of therapeutic options (e.g., end-of-life care), greater acceptance of system responsibility for medical errors and management mistakes, and a host of complexity-inspired management strategies. All of these arenas are ripe for exploration. A final, important consequence of greater appreciation of uncertainty is the performance and quality of worklife of the individual professional. Clinicians and managers who embrace uncertainty may have greater work satisfaction, longer job tenure, better career progression, and indeed a higher quality of life.

11.10 Summary and Conclusions

We have examined how uncertainty is perceived and managed at the patient-clinician and organizational levels of healthcare delivery. In both cases, the definition and management of uncertainty in healthcare reflects learned beliefs about the nature of the uncertainty and the role of clinicians and managers in reducing it. Those beliefs do not give much room to fundamental uncertainty, or to coexistence of uncertainty reduction with doubt and unsureness.

Social mechanisms to manage uncertainties in healthcare delivery have underemphasized the extent to which uncertainty is fundamental, both at the individual level and the organizational level. Further, educational programs and investment in knowledge development for uncertainty reduction have overemphasized the potential for knowledge to reduce uncertainty.

If healthcare delivery is to embrace uncertainty, clinicians, patients, and managers will need to be more realistic about the ability to reduce uncertainty through information. As a result, the roles of clinicians and managers will be less pressurized - less the roles of "superman" and "superwoman." Clinicians and managers will be partners and participants with patients, regulators, community representatives, and other stakeholders in an unfolding process that accepts fundamental uncertainty.

Modest changes toward a healthcare delivery system in which fundamental uncertainty is recognized and managed accordingly are in process. For example, acceptance of the fundamental uncertainty is implicit in the movement to reduce the blaming of individuals for medical errors, reflected in the phrase "to err is human" (Institute of Medicine 2000). Changing the mindset of current and future participants to a greater appreciation for and acceptance of fundamental uncertainty will move healthcare delivery forward at both the clinical and organizational levels.

References

Allen P (2000) Knowledge, ignorance, and learning. Emergence 2(4):78-103

Ashmos DP, Duchon D, McDaniel RR (2000) Organizational responses to complexity: The effect on organizational performance. Journal of Organizational Change Management 13:577-594

Ashmos DP, Duchon D, McDaniel RR, Huonker JW (2002) What a mess! Participation as a simple managerial rule to "complexify" organizations. Journal of Management Studies 39:189-206

Begun J, Kaissi A (2004) Uncertainty in healthcare environments: Myth or reality? Healthcare Management Review 29 (1):31-39

Begun J, Lippincott R (1993) Strategic adaptation in the health professions. Jossey-Bass, San Francisco

Begun J, Zimmerman B, Dooley K (2003) Health care organizations as complex adaptive systems. In: Mick SM, Advances in health care organization theory. Jossey-Bass, San Francisco

Boisot M, Child J (1999) Organizations as adaptive systems in complex environments: The case of China. Organization Science 10:237-252

Bonabeau E (2003) Don't trust your gut. Harvard Business Review 81(5):116-123

Clampitt PG, Dekoch RJ (2001) Embracing uncertainty: The essence of leadership. ME Sharpe, Armonk

Fiol CM, O'Connor EJ (2003) Waking up! Mindfulness in the face of bandwagons. Academy of Management Review 28:54-70

Friedman LH, Goes JB (2001) Why integrated health networks have failed. Frontiers of Healthcare Management 17(4):3-28

Glouberman S, Mintzberg H (2001a) Managing the care of health and the cure of disease - Part I: Differentiation. Health Care Management Review 26(1):56-69

Glouberman S, Mintzberg H (2001b) Managing the care of health and the cure of disease - Part II: Integration. Health Care Management Review 26(1):70-84

Griffith JR, Loebs SF, Dalston J (2001) Executive summary: Report of the national summit on the future of education and practice in health management and policy. Journal of Health Administration Education (19, Supplement):5-18

Institute of Medicine (2000) To err is human. National Academy Press, Washington, DC

Jamous H, Pelloille B (1970) Changes in the French university-hospital system. In: Jackson JA (ed) Professions and professionalization. Cambridge University Press, London

Kotter JP (1995) Leading change: Why transformation efforts fail. Harvard Business Review March/April:59-67

Luke RD, Begun JW (2001) Have integrated health networks failed in healthcare? Frontiers of Health Services Management 17(4):45-50

McDaniel RR (1997) Strategic leadership: A view from quantum and chaos theories. Health Care Management Review 22(1):21-37

McDaniel RR, Driebe DJ (2001) Complexity science and health care management. Advances in Health Care Management 2:11-36

Mechanic D (1996) Changing medical organization and the erosion of trust. Milbank Quarterly 74:171-189

Mirvis DM, Chang CF (1997) Managed care, managing uncertainty. Archives of Internal Medicine 157:385-388

Mishler EG (1997) The struggle between the voice of medicine and the voice of the lifeworld. In: Conrad P (ed) The sociology of health and illness: A critical approach, 6th edn. St. Martin's Press, New York, pp 295-307

Pascale RT, Millemann M, Gioja L (2000) Surfing the edge of chaos. Three Rivers Press, New York

Prigogine I (1997) The end of certainty. Free Press, New York

Schor R, Pilpel D, Benbassat J (2000) Tolerance of uncertainty of medical students and practicing physicians. Medical Care 38:272-280

Shortell SM, Kaluzny AD (2000) Health care management: Organization design and behavior, 4th edn. Delmar, Albany

Stacey RD (1992) Managing the unknowable. Jossey-Bass, San Francisco

Starr P (1982) The social transformation of American medicine. Basic Books, New York

Waitzkin H (1989) A critical theory of medical discourse: Ideology, social control, and the processing of social context in medical encounters. Journal of Health and Social Behavior 30:220-239

Weick KE, Sutcliffe KM (2001) Managing the unexpected. Jossey-Bass, San Francisco

Wilsford D (1994) Path dependency, or why history makes it difficult but not impossible to reform health care systems in a big way. Journal of Public Policy 14:251-283

Wingert TD (1996) The role of physician uncertainty in explaining variation in healthcare resource utilization. Ph.D. dissertation, University of Minnesota

Zimmerman B, Dooley K (2001) Mergers versus emergers: Rethinking structural change in health care systems. Emergence 3(4):65-82

12 Primary Care Practice: Uncertainty and Surprise

Benjamin F. Crabtree[1]

I will focus my comments on uncertainty and surprise in primary care practices. I am a medical anthropologist by training, and have been a full-time researcher in family medicine for close to twenty years. In this talk I want to look at primary care practices as complex systems, particularly taking the perspective of translating evidence into practice. I am going to discuss briefly the challenges we have in primary care, and in medicine in general, of translating new evidence into the everyday care of patients. To do this, I will look at two studies that we have conducted on family practices, then think about how practices can be best characterized as complex adaptive systems. Finally, I will focus on the implications of this portrayal for disseminating new knowledge into practice.

Let me give you a little background or context about primary care. As you probably know, most of the National Institute of Health (NIH) budget and a lot of the emphasis in recent biomedical research has been on tertiary care centers, hospitals, and high-tech interventions. Nevertheless, as reported in a study that was first done in 1961 by Kerr White, and later repeated in a *New England Journal of Medicine* report in 2001, a large proportion of medical care in the United States takes place in the primary care setting. In the 2001 report, Green and colleagues used recent national databases and found that for every thousand individuals in the country, about 800 of them report some kind of illness or injury in a given month. Of those, 327 consider seeking medical care and 217 actually visit a physician's office, with 113 of these going to a primary care clinician. So that's a fairly large percentage of 1,000 individuals who are going to a primary care office. Sixty-five go to a complimentary or alternative medicine provider; twenty-one get to an outpatient hospital; and fourteen get some home healthcare. Only thirteen go to an emergency room and only eight are hospitalized, with one (actually less than one) ending up in an academic medical center. So why do I think primary care is important? Because it is the foundation where most people are seen for most health care problems. So, if we are thinking about changing the health care system, my bias is that it should be done from the bottom up where most people receive care, at the primary care level.

I believe you are all familiar with the rapid and ongoing advances in technology and diagnostic capability that provide a theoretical capacity to significantly decrease morbidity mortality from a lot of common diseases. This ranges from diabetes to heart disease and different types of cancers. There have been many clini-

[1] Professor and Research Director, Department of Family Medicine, Robert Wood Johnson Medical School, University of Medicine and Dentistry of New Jersey, crabtrbf@-umdnj.edu.

cal trials and evidence-based reviews that are described in widely accepted practice-base guidelines about how to manage and prevent the common diseases. And yet, adherence to many of these guidelines has not met expectations. Why do we end up with this gap between the scientific evidence and what is actually going on in practice? One of the difficulties is that every day something new is coming in - there is a continual new onslaught of new technology; so practices and physicians need to learn every day.

Clinicians and practices have to be constantly changing; however, current models of organization and organizational change really limit how we can manage this anticipated uncertainty. What do we do right now? Well, Continuing Medical Education (CME) is a very common approach, which assumes that knowledge deficits can be countered and that this will translate into changes in practice. Another common approach is the development and dissemination of evidence-based guidelines; they are all over the web now. We send them out, but if you have seen some of these, they are very complex. For example, the cholesterol guidelines are almost two inches thick. Do you think a practicing physician is going to have time to go through two inches of a clinical guideline? The use of opinion leaders are often combined with chart reviews in order to provide clinicians and practices with performance feedback. In fact, the insurance companies do this all the time, but there is little evidence that this results in sustained change. Incentives and disincentives are approaches that actually work, particularly when a major incentive is whether you are paid or not. This is a fairly blunt instrument. Academic detailing is a technique that the pharmaceutical companies have perfected; although, they are now putting even more emphasis on consumer activation. Haven't you heard, "Allegra; see your doctor"? This is direct marketing. Finally there are different forms of office system innovations and there are different versions of Continuous Quality Improvement (CQI).

The problem is that while there is some evidence that each of these change strategies work, the reality is that they don't work in a consistent way. It's not like there is a magic bullet here that you can just say, "we are going to do a CME and it's going to make a difference." Generally, we have to do combinations of these and still only get limited results. So what is the problem?

Each of the common dissemination strategies assumes that an intervention will result in a proportional response. We now know that is not a valid assumption. Sometimes you have a very minor intervention that has a big change, and sometimes you can hit people over the head and it doesn't seem to make any difference.

I will now turn to two studies of primary care practices on which I was a participating investigator that led us to explore applications of complexity science to primary care practice change. The first is the Direct Observation of Primary Care (DOPC) study. This was funded by the National Cancer Institute and conducted in Northeast Ohio. The second is the Prevention and Competing Demands in Primary Care (P&CD) study that was funded by the Agency for Healthcare Research and Quality and conducted in Nebraska. The DOPC study was a cross-sectional study of 84 practices with 138 physicians in those practices. Almost 4500 patients were directly observed using a tool called the Davis Observation Code. Nurses observed care in the examination room and were prompted by a little earplug every fifteen seconds - that is, every fifteen seconds a recording says "observe, record; observe, record." The observers have twenty-two variables that they can check off in each of these segments, ranging from history taking to counseling. They can mark off multiple items in each time segment, so for every fifteen seconds there are twenty-two variables. Since the average encounter is ten minutes, well, you can do the math on the size of the data set. The DOPC thus provides a very detailed quantitative data about patient care. In addition to the Davis Observation Code, a medical records review was done for every patient that was observed, along with an exit questionnaire and an abstraction of billing data. At the end of the day, the nurse observers dictated a summary of what they observed and how they felt about it.

The other study, the Prevention and Competing Demands (P&CD), is almost the opposite. It was designed as a follow up of the DOPC study in order to look really in depth how the practice and the people in the practice worked together. We examined eighteen practices; including some that were doing really well in delivering preventive services and others doing poorly. Observers spent weeks and weeks observing, taking detailed notes about how the organization runs, and talking to the people. They observed thirty encounters with each clinician and dictated fieldnotes in the form of a chronology of what went on during the encounter. The P&CD study generated about 20,000 pages of text for analysis.

Over the past six or seven years we have published more than fifty peer-reviewed manuscripts from these two studies. The results of these studies set up why we think complexity science is important. In looking at these studies, one of the things that becomes apparent is that there is a lot of variation - a lot of good variation, as well as a lot of not so good variation. For example, physicians are prioritizing care, thinking of care in an ongoing, continuous relationship with a patient. They don't do everything in one day; but actually do it over time. Physicians are also using the risk factors to tailor care. For example, if someone has bronchitis, this can really trigger a lot more about smoking or some other topics. They are also looking for teachable moments; times, opportunities when there is a good time to reach patients. So there is a lot of good variation.

There is also a lot of problematic variation. For example, the over-prescription of antibiotics is very common. We looked at about 800 encounters where the patient had an upper-respiratory infection. A very high percentage of these patients got an antibiotic, and most of these were inappropriate according to Center for Desease Control (CDC) guidelines. When we started looking at what goes on in these visits, patients were really pressuring physicians, often indirectly. For example, there is the Disney Land story that goes something like, "I have been saving for months to go to Disney Land. We are leaving tomorrow; you have to do something." Other patients would present with a severity story, saying "I'm really sick, you have to do something." That will get you an antibiotic almost every time. So the patients are really a big part of shaping the care in the practice.

But physicians contribute to this variation as well. In looking at smoking, management of depression, and other common healthcare issues, it was readily apparent that some physicians function better at providing services than others. For example, some physicians really have their antenna up for depression or smoking and are providing services on a constant basis. Others are not doing it as well.

We also see a lot of variation at the practice level. One of the things we looked at is how staff are used. Perhaps somewhat surprising to some, there are family practices that have no registered nurses, often referred to as an RN. Some practices may have one RN who is only there as a clinical manager, but many use very low skilled, relative to medical training, medical assistant's (MA's) that may have as little as six months of formal training. Nursing staff tend to take weights and blood pressures and give injections. But often that's about it. Even in places that had RNs, it was common for their role to be limited to taking blood pressures and weights and putting the people in the rooms anyway.

Geographic location also creates variation. Urban practices are very different from rural practices; rural practices are very different from suburban practices. Geography actually matters, as does characteristics of the community being served.

The larger health care system also creates variation and surprise. Twenty-four percent of patients in managed care had to change their physician in a two-year period in the DOPC study. This creates a lot of discontinuity in the care of patients that makes guidelines difficult to implement.

Finally, the visits themselves are very complex. Primary care visits take an average of ten minutes, but within this time, there is a lot of different kinds of problems addressed; and usually not one problem; but three, four, five, and six problems. These are being prioritized. And then in eighteen percent of the visits, there is not just one patient; there is more than one person. For example, this might be a mother and her child, but both are treated, and often while only billing for one of them. Then there are patients who come in who are depressed. And this really impacts on the visit. Even if the clinician doesn't do anything, it makes the visit about two minutes longer; and it goes up to three minutes longer if the clinician actually talks about the depression. This has a big impact.

So, how do we make sense of all this variation and complexity? How can sensible change take place in these contexts? Based on insights from the DOPC and P&CD studies, we started working on a model that incorporated characteristics of patients, clinicians, the clinical encounter, the practice, the community and the larger health system. There is a lot of complex interactions going on in practices and we struggled to make sense of it all. We realized that practices are faced with the need for ongoing change in response to new technology and innovations, and felt we needed to have a model that could help us understand multiple characteristics simultaneously since these were obviously not independent. So we started our early explorations with Dr. Reuben McDaniel looking at practices as complex systems. We could see that practices co-evolved with their local communities to fit the needs of their particular patients. Thus, it is not surprising to see variation between an urban practice and a suburban practice, because they have very different kinds of patients, disease patterns, and staffing patterns. When we started looking at what makes a difference, it became apparent that how the practice got started, the initial conditions, are very important. Key questions became, Who are these people? What is their training and how do they work together? What are the relationships? What is that local fitness landscape? What are the regional and global influences?

Complexity science offered an obvious framework for understanding the variation and diversity and for addressing these questions. Three properties of complex systems of particular relevance are the properties of self-organization, co-evolution, and emergence. We found that by looking at the data we could see evidence for each of these as these organizations evolved. We could see developments in local landscape that impacted a practice and could see practices co-evolving as other systems around them co-evolved. We could see the emergence of whole systems and that it was not possible to tease apart different components of the practice like most of the researchers want to do.

Here are a couple of quick examples to illustrate. Everyone can probably resonate with these fairly easily. A suburban practice, Franchise Family Practice, is captured by the franchise metaphor. A hospital system created the practice in a rapidly expanding suburban community in order to focus on peoples' expectations - getting people in and out efficiently. The whole practice organizes towards that goal and maintaining a profit. The receptionist had a script on how to answer the phone; how to talk to people. They always got your insurance card when you came in. Everything went like clockwork, getting you in and out. You always got billed and most people were quite pleased with the services the practice provided.

Contrast this with Dusty Garden Family Practice, a different metaphor. Dusty Garden is an inner-city practice serving a local indigent population. The practice had a vision to care for the under-served. Where they interested in getting people in and out efficiently? No, they were really interested in relationships. They hired the staff from the local community, which had a lot of minority patients. As the practice evolved, it was a very different place than Franchise Family Practice. Both Franchise and Dusty Garden co-evolved with very different systems to create very different organizations, each adapted to a particular fitness landscape.

If this model of organizational self-organization, co-evolution, and emergence is accurate, then there are implications for interventions designed to disseminate new technology and ideas (and to enhance existing care processes). Based on our emerging understanding of complexity science, we developed a clinical trial called the Study to Enhance Prevention by Understanding Practice (STEP-UP). This study was also funded by the National Cancer Institute. We randomized eighty practices in Ohio, and used an initial assessment of the practice to provide insights for tailoring feedback to the practice. Briefly, a facilitator spent just two to five days in the practice observing what was going on and used some structured checklists and some physician and staff surveys. We developed a genogram to give us a sense of how they were organized. A genogram is essentially an expansion of an organization chart that shows functional relationships and communication patterns. And then we had a tool kit of the kinds of options for change they could use. Based on our talking with the people, we understand who works with whom; who communicates with whom. Lightening bolts are conflict-type of relationships; double lines are like a healthy relationship. A single line is an okay; it's a positive relationship. If we had triple, it would be in an enmeshed relationship. The dotted lines mean people are isolated from the other groups. So you have a group. This was the front office staff that was isolated from the rest of the people. We had ghosts in there; people who were gone, but were still influencing. Those are the ones with "X's" on them. So, a genogram is actually a picture of a practice that gives a sense for who are these people and how they work together. Based on the assessment, we would go back to the practice, show them their genogram, give them their report, and then negotiate the intervention. So, how did we do?

The STEP-UP study was one of our earlier applications of complexity science to facilitating change in practices. Before the intervention, the control group was actually somewhat better than the intervention group in the delivery of global preventive services. Within six months the intervention and control groups were the same and within twelve months the intervention group was significantly better. This is for a full range of preventive health services. We did a delayed intervention in the control group, so that's why we don't have comparison data going forward. But what is interesting is that the intervention group didn't go down at the completion of the study. If you look in the literature, it is very common that after the intervention is over, practices go back to baseline fairly quickly. We now have four years of data and it appears that these practices have made permanent changes; somehow, or the other, there is something happening in those practices that created sustained change. We believe that one of the implications from this STEP-UP study is that we need to tailor interventions to fit the local landscape. Probably we also need to facilitate this change process in order to identify a practices' current capacity for change.

In our latest work, we are investigating approaches for stimulating self-reflective, ongoing learning in practices. Based on an ongoing analysis of the qualitative process data collected as part of the STEP-UP study, we have developed a refined model of organizational change. This analysis, which was also funded by the National Cancer Institute, focused on finding out what about the intervention worked and why it worked in some places and not in others. From this we developed the ULTRA study, which stands for Using Learning Teams for Reflective Adaptation. The ULTRA study is funded by the National Heart, Lung, Blood Institute and started last fall. In this study we do an initial assessment of the organization, using what we call the Multimethod Assessment Process (MAP). The Reflective Adaptation Process (RAP) is an iterative team building process that combines assessment feedback with the facilitation of learning teams in the practice. In this study, we are doing sixty practices in New Jersey and Pennsylvania, again randomizing practices into an intervention and control groups. We do a two-week assessment, followed by three to six months of facilitated follow-up. The hypothesis is that by changing overall practice processes, it is possible to simultaneously impact a wide range of outcomes including smoking, hyperlipademia, hypertension, diabetes, asthma, practice culture, and other things. We suspect we will be able to change all of these without focusing directly on any single one based on our understanding of complexity science. That is, we are changing the relationships of the agents and making learning teams of physicians, nurses, and front-office staff. As part of the intervention process we are really working with them weekly; an hour a week we sit down with them, so the practices actually have to carve out an hour per week and work together. We have our facilitator there to help the practices in changing the relationships. We think that by changing those patterns of relationships among agents that the whole system is going to change. We don't even have to talk to them about asthma, because they want to do that anyway.

In conclusion, we are really looking at practices as complex systems and taking advantage of insights from complexity science. Finally, this is just an overview of the intervention, but we want to give one final caution that is attributed to Yogi Berra, "in theory, there is no difference in theory and practice, and in practice there is." That's it.

13 Medical Errors and Microsystems: The Best Things Cannot Be Told

John C. Peirce[1]

13.1 Introduction

The best things cannot be told
Because they transcend all thought.
The second best are misunderstood
Because those are the thoughts that refer
To that which cannot be thought about,
And so we're stuck with our thoughts.
The third best are what we talk about.
Joseph Campbell

13.1.1 Spontaneous Intuitive Behavior

We cannot say all the things that we do when we go about the spontaneous intuitive behavior that constitutes everyday life - whether it's driving an automobile, tying one's shoes, or speaking in one's native language in a phonetic, syntactic and idiomatic style unique to our culture and area of the country (Schön 1987; Groopman 2000). In every circumstance, one cannot cite the rules for all that we do. Yet as a rule, each person *knows how* to do these activities. *The knowing is in the doing*. This is called tacit knowledge - that which is not spoken. It comprises over 99 percent of what we do day in and day out. Neurobiologically, this activity is processed outside of the conscious cortex (Damasio 1999).

We consciously attend to something when we are *surprised*, either pleasantly when the outcome of our actions is promising - or troubled when the outcome is unwanted. These internal feelings are inextricably woven with thoughts, and these parallel streams of thinking and feeling are at the core of judgment (Damasio 1995; Damasio 2003; Schön 1983). We act upon troubling feelings by changing direction (reflecting) during the activity - or by reflecting after the fact and reorganizing our thoughts and subsequent action. Reflection is at the core of how professionals learn (Schön 1987). By daily practice, people aspiring to become professionals convert their behavior from that prompted by conscious deliberate rules to that which is spontaneous and intuitive bypassing the conscious mind. In so doing, their performance moves from being laborious and protracted to fluent, spontaneous and timely in which they can call forth a large inventory of actions to meet the situation at hand (Dreyfus and Dreyfus 1986).

[1] Former Chief Academic Officer, Banner Good Samaritan Medical Center, Phoenix, jcpeirce@comcast.net.

But all too often, our efforts in this regard go for naught; we reach the limits of our individual capabilities. When this happens, we begin to talk with others in order to make sense of what is happening that causes us distress (Weick 1995). These are the seeds of organizing - seeking out different experiences to enrich our understanding and soliciting different courses of action to see what works better. Ongoing conversation is at the heart of how cultures change (Shaw 2002). It is well to remember, though, that

> ...organizing is never very tidy or foresightful despite the necessity of its practitioners to make it appear otherwise. Efforts to maintain the illusion that organizations are rational and orderly in the interest of legitimacy are costly and futile. They consume enormous energy and undermine self-acceptance when managers hold themselves to standards of prescience that are unattainable. (Weick 2001)

13.2 Cultures Co-evolve

As to the future, our task is not to foresee it but to enable it.
Antoine St. Exupery

Conversations are the major way by which we organize. When making sense of important issues, we build new conceptual models to better explain what is happening and find new ways that work better. We come to believe in new ways of explaining and doing things through collective learning. While lectures, workshops and other formal educational exercises are useful, informal conversations are the major way by which our beliefs change. In these small personal discussions one person can challenge another to determine whether the other really believes in this new way and whether it actually works. It is through this give and take that cultures co-evolve - especially the basic assumptions that are the tacit framework for the culture (Schein 1992).

Having laid the groundwork of spontaneous intuitive behavior and the co-evolution of culture, we will proceed to two actual cases of medical errors.

13.2.1 Two Cases

First Case

A 45 year old male was found on the ground at a bus stop. He had water bottles tightly strapped to his thighs, abdomen and penis to "ward off evil spirits." He appeared to be acutely ill and was transported by the emergency medical transport (EMT) unit to the emergency department of a large teaching hospital. The emergency personnel found him to be frankly delusional with a blood pressure of 85/55, heart rate of 135 per minute and a large swollen and purple to black discoloration of his scrotum and penis with blackened skin on a red base in the left lower quadrant of his abdomen extending to the upper lateral aspect of this left anterior thigh just below the inguinal fold. He had a blood glucose of 35 mg/dl, serum creatinine of 2.7 mg/dl, bicarbonate of 9 mM/L, and a white blood cell count of 1,900/cubic mm. [For those with little or no clinical experience, this man was seriously ill and moribund or close to death.]

Within several hours he was admitted to the intensive care unit and was seen by an attending physician. A dopamine drip was ordered to raise his blood pressure. Penicillin, gentamicin, clindamycin, metronidazole and cephepime [all antibiotics] were ordered, and a STAT urology consult was ordered. Within two hours the patient coded and died.

At a Morbidity and Mortality Conference the following problems were identified: 1) The patient had necrotizing fasciitis with septic shock requiring immediate surgical removal of dead tissue and management of his septic shock, 2) A small peripheral vein was all that was available for the dopamine drip, 3) The unit clerk called and talked only to the office personnel in the urologist's office for the consult, 4) The attending physician was in his private office 5 minutes away when the patient coded, and 5) An in-house critical care specialist was discouraged from seeing the patient.

Taking into consideration that judgment depends upon intertwined streams of feelings and thought, what are your feelings and thoughts, and what you do think ought to be done now?

Second Case

A 67 year old woman was admitted to another teaching hospital for cerebral angiography. Several months earlier she had fallen striking her head. An MRI at that time revealed two large cerebral aneurysms. During the angiography, one the aneurysms was successfully embolized. The second aneurysm was deemed more amenable to surgical therapy, which would occur at a date in the near future. After the procedure she was transferred to an oncology floor for discharge on the following day rather than to her original bed on the telemetry floor.

The following morning one of the oncology floor nurses took the patient to the electrophysiology laboratory for a procedure despite the patient's having raised objections to both the nurse and the physician who called her to explain the procedure to her. About one hour into the procedure, it became apparent that she was the wrong patient, and the procedure was aborted. She returned to her room, and the physician explained that a mistake had been made. She was discharged the following day with no further problems.

A root cause analysis revealed that 17 interaction errors occurred involving 14 people (including seven RNs, three physicians and the patient) in five microsystems (Chasin and Becher 2002) (see Table 13.1).

Table 13.1. Interaction errors in this case

Interaction errors in this case[a]
1. An unidentified person on the telemetry floor misdirected RN1 by saying "patient Morrison" was not on floor (when she was) and by saying that she had been transferred to oncology (6:15 a.m.).
2. An unidentified person on the oncology floor misdirected RN1 by saying patient she sought (Ms. Morrison) was on the floor when she was not. (6:20 a.m.).
3. An unidentified person on the oncology floor told RN2 to bring her patient (Ms. Morris, the wrong patient) to the electrophysiology laboratory (6:30 a.m.).
4. RN2 took her patient to the electrophysiology laboratory despite a) the patient's objections, b) a lack of consent form and order in the chart, and c) lack of knowledge on her own part or that of her charge nurse that the procedure was planned. (6:45 a.m.)
5. RN1 failed to verify patient's identity against the electrophysiology laboratory schedule when the patient arrived in the electrophysiology laboratory (6:45 a.m.).
6. RN1 failed to recognize the significance of Ms. Morris's objections to undergoing the procedure (6:45 a.m.),
7. The electrophysiology attending physician failed to verify Ms. Morris's identity when he spoke to her by telephone, and he failed to understand the basis of her objections to the procedure (6:45 a.m.).
8. RN1 failed to appreciate the significance of the lack of an executed consent in the chart, especially given that the electrophysiology schedule stated that the correct patient (Ms. Morrison) had signed the form (6:45 to 7:00 a.m.).
9. The electrophysiology fellow failed to verify the patient's identity, failed to recognize the significance of the lack of pertinent clinical information on her chart, and failed to obtain a consent that was informed (7:00 to 7:15 a.m.).
10. The electrophysiology charge nurse failed to verify the patient's identity (7:10 a.m.).
11. RN3 failed to verify the patient's identity (7:15 to 7:30 a.m.).
12. The neurosurgery resident did not persist to obtain a satisfactory answer to his question as to why his patient was undergoing a procedure about which he had not been informed (7:30 a.m.).
13. RN4 failed to verify patient's identity (8:00 a.m.).
14. The electrophysiology attending physician failed a second time to verify patient's identity when he did not introduce himself to Ms. Morris at the beginning of the procedure (8:00 a.m.).
15. The electrophysiology fellow disregarded the fresh groin wound from Ms. Morris's cerebral angiogram the day before and started the electrophysiology procedure on the opposite side (8:00 a.m.).
16. A telemetry nurse (RN5) and two electrophysiology nurses (RN3 and RN4) failed to verify the identity of the patients they discussed on the telephone (8:30 to 8:45 a.m.).
17. The electrophysiology charge nurse failed to persist in obtaining a satisfactory answer to her question of why no patient with the name Joan Morris appeared on the electrophysiology schedule (8:30 to 8:45 a.m.).

[a]Times in parentheses refer to "Chronology of Events" section of the text.

As in the first case, what are your feelings and thoughts, and what you do think ought to be done now? Compare these with those you had in the first case.

13.2.2 Discussion

If you are like most people, the natural tendency is to find fault with one or several of three health care workers in the first case, particularly the attending physician, and to a lesser extent the ICU nurse or ward clerk. One has difficulty carrying individual faultfinding over to the second case, however. There are too many people - including the patient, and there are too many interactions. *The errors in this case exemplify a tacit cultural pattern that transcends any given individual.* Inattention to identity checks has become more a group responsibility than an individual responsibility.

13.3 Complex Adaptive Systems

This case also exemplifies two of the characteristics of a complex adaptive system: many people (agents) and many interactions that are nonlinear, resulting in patterns of self-organization, emergence and co-evolution. I use the word, "system," to mean a pattern of interactions that persist yet evolve. By the nonlinear I mean that if just one of the 17 involved persons - and it doesn't matter who - exercised sufficient assertive diligence to ensure that the patient's identification matched that of the person having the electrophysiology procedure, i.e., broken the cultural pattern, this would have prevented the error before it occurred. A little intervention would have yielded a large effect. The opposite is also true. Having a large number of health care professionals go through substantial time and effort to create a policy and procedure would have produced very little effect if a substantial number of health care workers did not change their spontaneous intuitive behavior, thus developing a new cultural pattern. Lastly, this error was unpredictable at this time and with this patient because the number of people and interactions were too great to model this with any degree of specificity. It surprised the health care workers involved. What we can say is that errors of identity will occur with increased frequency with cultural norms found in this case as compared to other groups that are intuitively compulsive about identity checks. With this in mind, return to the first case and make sense of it using the ideas just discussed.

13.4 Micro-systems - as well as Meso-systems and Macro-systems

The second case also exemplifies micro-systems, a concept introduced by Nelson and Batalden (2000). In this situation the telemetry floor, interventional radiology suite, oncology floor, electrophysiology laboratory and physician's office constitute five separate micro-systems. In the Nelson and Batalden (2000) model, the core of the micro-system is the patient with a physician and group of health care professionals (usually led by a nurse team leader) interacting with a patient and population of similar patients. *Since the micro-system is the only unit where the patient and her or his family interact with the larger health care system, this is the* ***only place*** *where the intuitive expertise of the patient and their family can be effectively communicated and dealt with* (Groopman 2000). Kleinman, Eisenberg and Good (1978) distinguish between illness and disease wherein illness is how being sick affects the patient and his or her family including how they make sense of being sick. Disease is how we health professionals recast illness into abnormalities of structure (pathology) and function (pathophysiology) of the body along with causal factors (etiology). While thinking in a disease perspective gives us tremendous power in addressing these problems, we run the risk of building distance between ourselves and our patients by becoming too abstract and diminishing personal interactions that go to the heart of patient's concerns. Information systems that support the actions of the caregivers, patient and family are an essential part of a good micro-system. Finally, a micro-system has support personnel and manager with equipment and clinical environment in which the entire unit operates.

This model was adapted from studies by Quinn (1992) who showed that those large enterprises that are more profitable provide information systems that better support the actions of the personnel (and the consumer) in the small functional units where the people who receive services from the enterprise interact with it. This also is supported by work of the Gallup Organization (Buckingham and Coffman 1999) who derived 12 questions that determined how well employees are engaged with their companies. Positive answers to these questions were highly linked with increased productivity, profitability, employee retention and customer satisfaction as shown in Table 13.2.

Table 13.2. Twelve questions that predict increased productivity, profitability, employee retention and customer satisfaction

Questions that predict increased productivity, profitability, employee retention and customer satisfaction
1. Do I know what is expected of me?
2. Do I have the materials and equipment I need to do my work right?
3. At work, do I have the opportunity to do what I do best every day?
4. In the last seven days, have I received recognition or praise for doing good work?
5. Does my superior, or someone at work, seem to care about me as a person?
6. Is there someone at work who encourages my development?
7. At work, do my opinions seem to count?
8. Does the mission/purpose of my company make me feel my job is important?
9. Are my co-workers committed to doing quality work?
10. Do I have a best friend at work?
11. In the last six months, has someone talked with me about my progress?
12. This last year, have I had opportunities at work to learn and grow?

Ten of twelve questions were linked to *productivity*, and eight of twelve were linked to *profitability*.
One, two, three, five and seven were linked to employee retention. These are most strongly linked to the *employee retention*, leading the researchers to conclude that employees leave managers, not companies.
One, two, three, four, five and six are linked to *customer satisfaction*.

Micro-systems exist within meso-systems of health care (meso = intermediate), such as hospitals, medical centers, or multi-institutional health systems. Meso-systems with their micro-systems are part of macro-systems, i.e., interstate health plans, governmental and state health systems. These relationships are a matter of policy or are contractual and involve the flow of funds.

13.5 Evolution of Complexity in Medicine During the Last Half of the 20th Century

Fifty years ago, only a physician's office of the ones mentioned above existed as a micro-system. The other seven micro-systems in our two cases, emergency medical transport, emergency department, intensive care unit, telemetry floor, interventional radiology suite, oncology floor, electrophysiology laboratory, had yet to emerge. As noted in Table 13.3, the vast majority of physicians in the first half of the 20th century were in general practice (Rothstein 1987), most of whom had only one year of a rotating internship. They used a hospital emergency room as an extension of their private office where they could repair a laceration, remove a foreign body from an eye, set a fracture or examine an acute abdomen for the possibility of appendicitis or an acute gall bladder. They used to admit patients to the hospital who were more acutely ill and more complicated so they could receive nursing care, have available operating rooms, labor and delivery facilities, a pediatric floor for the care of children, radiology and clinical laboratory services and have a place where a specialist could see them. Only in large city-county hospitals were there 24 hour, 7 day a week coverage of emergency wards and the hospital by residents. During the 1950's, the number of general practitioners dropped precipitously so that by 1961 they were well below 50 percent. The increase in physicians practicing internal medicine and pediatrics kept primary care just over 50 percent, but by 1970 the sum of all primary care physicians had become a minority where it would remain. By 1970, "other" active practicing physicians had risen to over a quarter of all actively practicing physicians showing the emerging number and diversity of specialists. Fifty years ago, health care had far less capability in caring for a variety of illnesses, and it was far less complex.

Table 13.3. Percent physicians in active practice during the 20th century

	1923	1931	1940	1949	1961	1970	1983
GP/FP	89.4	83.5	77.7	67.3	38.4	18.6	13.9
Int Med	1.3	2.7	3.9	6.0	12.2	13.5	17.8
Peds	0.5	1.0	1.5	2.3	4.8	5.8	7.0
Med Spec	0.5	0.7	1.0	1.9	2.6	5.6	6.4
Gen Surg	2.2	2.9	4.0	5.2	9.3	9.6	8.0
OBG	0.5	0.9	1.5	2.6	5.6	6.1	6.4
Surg Spec	4.0	5.7	6.3	7.6	10.8	12.0	12.7
Other	1.5	2.5	4.1	7.1	16.2	27.3	28.6

The energy and resources that brought about this accelerated change came from World War II. In a matter of a year or two, the US mounted a war effort and brought forth victory within four years not only in North Africa and Western Europe but also in the Pacific. Prior to the war, penicillin was known but it was produced in minute quantities at great cost. Within a year or so into the war penicillin was being turned out in 15,000 gallon tanks, and by the end of the war, it was available for both military and civilian use. After the war federal research dollars came forth in increasing amounts, health insurance took hold, and for a while, the GI Bill supported physicians wanting to obtain specialty residency training. The general practitioner began to disappear; specialists in medicine and surgery began to care for the severely ill patient in intensive care units, an offspring of the post-anesthetic recovery unit. The transistor led to telemetry and telecommunication that helped foster emergency medical transport and telemetry units. The emergency room was transformed from a single room to a department in which physicians were on site 24 hours a day, 7 days a week. A greatly expanding menu of anti-cancer medications in a variety of combinations fostered inpatient oncology floors. Increased imaging capabilities with intravascular and intraluminal catheters fostered the development of interventional radiology. Cardiac catheterization, which was primarily a research tool in the 1950's, advanced so that by the 1990's cardiologists could map the intra-cavitary conduction system of the heart leading to the subspecialty of electrophysiology. Each of these disciplines developed its own specialty or sub-specialty with nursing, technical and administrative support - often as a service line. As shown in Table 13.4, the number of health care professionals needing to support physicians increased four orders of magnitude from the beginning of the 20th century to the 1980's, and patients in the ambulatory setting are increasingly seeing non-physician clinicians (Aiken 2003; Druss, Marcus, Olfson, Tanelian and Pincus 2003).

Table 13.4. Decrease in the proportion of physicians among health care workers during the 20th century

Time	Proportion of physicians among health care workers
Turn of the 20th Century	1 in 3
1980's	1 in 16
Time	**Proportion of ambulatory patients seeing a non-physician clinician**
1987	31 %
1997	36 %

These emerging programs occurred not by a grand plan and engineering design but through multiple self-organizing efforts spawned by research funding, an increasingly rich national and international network of meetings and communication among researchers, and the ability to implement one's programs with specialists funded by health insurance. This also spawned a co-evolving service line administrative infrastructure. This exemplifies the final three characteristics of a complex adaptive system: self-organization (coming from an energy source), emergent phenomena and co-evolution; this completes Table 13.2.

To add to the complexity, the proportion of women graduating from medical school has increased from under 10 percent in the early 1970's to 49 percent in 2002 as shown in Figure 13.1 (Medical Education, 2002). This was independent of the progress in medicine, coming about through the various forces of the women's movement of the 1960's and beyond.

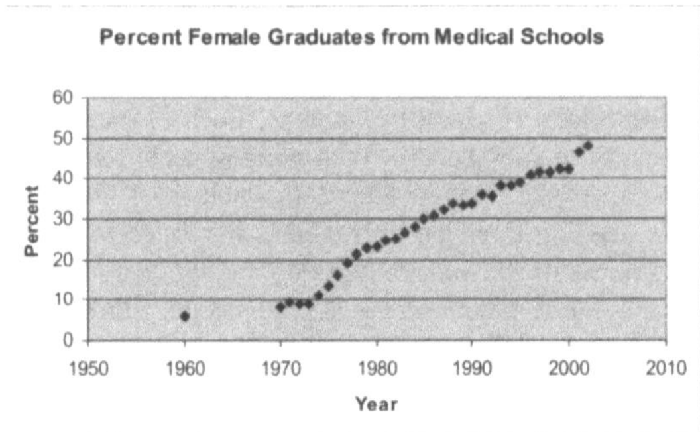

Fig. 13.1. Women in medicine data

What has not evolved has been an adequate funding system. We still have a health care funding system that originated when general practitioners were the great majority of physicians. Now we have no general practitioners and primary care physicians are in the minority. Health care micro-systems exist in ambulatory settings with physicians, in hospitals or in freestanding facilities such as endoscopy facilities. A conflict of interest often exists between the hospital or facility and the physicians. A dedicated funding stream to physicians has promoted a culture of individual responsibility often to the exclusion of collective responsibility necessary to the effective operation of micro-systems. Physicians have failed to discern the paradoxical nature of individual and collective responsibility, the latter having a "both-and" character, not an "either-or" one.

13.6 Where Do We Go From Here?

Shaw (2002) argues that cultural change happens primarily through conversation, noting that the word comes from "con," which means "with" and "vers" meaning "turning" - making conversation that which we do when we *turn with* each other. Progress leads to emergent phenomena, which require assimilation into what already exists - resulting in confusion. This leaves us with the necessity of making sense of what is happening and how to deal with it. Misunderstanding, paradox, ambiguity and fundamental uncertainty are naturally part of this confusion. It does us well to remember that in our conversations, "the best things cannot be told" and "the second best are misunderstood." The old perspectives, since they have worked best in the past, tend to dominate. Other points of view will emerge if there is sufficient diversity of experience among those engaged in the conversations. Conflict is usual and conversations are messy. An important way of facilitating newer ways of thinking is *being able to ask the right questions*.

In elaborating how to ask the right questions, Glouberman and Zimmerman (2002) compare and contrast complicated versus complex systems. Sending a manned space vehicle (MSV) to the moon requires a complicated system whereas raising a child is a complex one. Protocols are necessary and critical for sending a MSV to the moon because MSVs are similar in many aspects. Even so, expertise is still required where intuitive experience can be called forth when needed. Success in sending one MSV to the moon increases the likelihood of the next one being successful. There is a high degree of certainty about the outcome leading to optimism. On the other hand, protocols and formulae have very limited application (an immunization schedule would be one example) in raising a child. Expertise can contribute but is neither necessary nor sufficient to ensure success because every child is unique and must be understood as an individual - not as a standardized person. We are always uncertain of the outcome nonetheless we can remain optimistic if we pay sufficient attention to the child and put her or his best interests above our own.

The authors contend that our problems in health care are that we treat everything as a complicated system ignoring that most are complex. This leads us to a profusion of laws, rules and regulations, policies and procedures. Since all of these, by their very nature are context-free, they assume that all micro-systems are similar. A family practice office in Grand Rapids, Michigan that deals with patients who are middle class and come predominantly from Dutch and Polish backgrounds is assumed to be similar to a family practice clinic in Los Angeles whose population is predominantly lower middle class and come from Southeast Asia and Mexico. The same can be said about radiology departments in Bronx, New York in contrast to those in Greeley, Colorado. This does not make sense because customs, culture, politics, available resources and history will make each locale and their micro-systems unique. This makes asking the right questions essential.

Table 13.5 contrasts questions that assume that a complicated solution is required - wherein the responsibility rests primarily with the meso-system or the macro-system (left hand column) with those in the right hand column where a complex solution is required and responsibility rests primarily with the micro-system and their self-organizing capabilities. Context-free rules can never address how being sick affects the individual patient and their families. Dealing with illness, in contrast to disease (Buckingham and Coffman 1999), can only be addressed by the micro-system, wherein healing, therapeutic relationships are formed. Context-free rules are sterile and cannot take into account individual needs, values and desires; just ask any patient who has had to deal with the protocols of a health plan.

Table 13.5. Questions contrasting complicated versus complex situation

Complicated questions Responsibility with meso- or macro-systems	Complex questions Responsibility with micro-systems
What sort of policies and procedures do we need to put in place to reduce errors?	How can we do things differently to reduce errors in our micro-system?
What sort of program do we need to develop in our hospital, health care system or health plan or governmental agency to ensure that everyone is "on board" regarding safety; especially to ensure good communications among all physicians, nurses and other health professionals?	What sort of methods of communication within our micro-system - and between us and other micro-systems we use for our patients - will engender safer practices? And how can we measure this?
With our limited resources, what should we eliminate to develop our new safety program?	In our micro-system, how can we continuously develop safer practices using the resources we now have? (reducing waste that will free up time for more effective practices)
How can we sanction those programs that are not in compliance with our rules and regulations regarding safety?	How can we use playfulness, humor and celebration to engage one another within and between our micro-systems to bring about safer practices?
What protocols do we need to put in place to contain costs?	What can we do differently that will lower our cost per encounter – and maintain or improve the quality of the services we deliver?

Weick and colleagues argue that too many rules and regulations, policies and procedures have an opposite effect of what they were intended to accomplish (Weick, Sutcliff and Obstfeld 1999; Weick and Sutcliff 2001). They refer to this as "over specification," that which is often called bureaucratic red tape. In their studies of high reliability organizations, these researchers found that organizations with under specification are associated with a greater use of available expertise needed to address present problems than those organizations that are overburdened with policies and procedures that inhibit the use of available expertise; the latter results in a higher error rate.

Meso- and macro-systems are essential and critical for providing necessary infrastructure services such as contracting, financial management according to generally accepted accounting principles, policies and procedures in human resources that ensure compliance with employment law, and so forth. So conversations about complex safety issues require including those in the meso - and sometimes the macro-systems - making this a paradox that needs managing. Nonetheless, the primary responsibility for errors rests with the micro-system.

The messiness and difficulty of these conversations constitute a crucible where new leadership may emerge. There are those who will intuitively understand paradox and be able to shift the conversation from an either-or mode to a both-and one. This may be shown in the ability to work through the necessity of both individual and collective responsibility and enabling funding patterns that will diminish funding bias. These emerging leaders will also show a capability of staying with the conversation in spite of its ambiguity and fundamental uncertainty to work through misunderstandings. This allows for differences to be resolved through ongoing experiences, which permit different points of view to converge through conversation.

In the first case, a good question might be: How can we ensure that a physician skilled in managing this complex and unstable clinical problem is in attendance until the patient stabilizes? In the second case a good question might be: How can we become compulsive about identity checks within and among our several micro-systems.

Through it all, we return to the wisdom of T. S. Eliot:

We shall not cease from exploration
And the end of all our exploring
Will be to arrive where we started
And know the place for the first time.
T.S. Eliot (Little Gidding, Four Quartets)

References

Aiken LH. (2003) Achieving an interdisciplinary workforce. N Engl J Med 348:164-166

Buckingham M, Coffman C (1999) First break all the rules: What the world's greatest managers do differently. Simon and Schuster, New York

Chasin MR, Becher EC (2002) The wrong patient. Ann Intern Med 136:826-833

Joseph Campbell and the power of myth with Bill Moyers, 2. The message of myth. DVD. Mystic Fire Video, http://www.mysticfire.com, 1988

Damasio A (1999) The feeling of what happens: Body and emotion in the making of consciousness. Harcourt Brace and Company, New York

Damasio A (1999) Descartes' error: Emotion, reason and the human brain. Avon Hearst, New York

Damasio A (2003) Looking for Spinoza: Joy, sorrow and the feeling brain. Harcourt Brace and Company, New York

Dreyfus HL, Dreyfus SE (1986) Mind over machine: The power of human intuition and expertise in the era of the computer. The Free Press, New York

Druss BG, Marcus SC, Olfson M, Tanelian T, Pincus HA (2003) Trends in care by nonphysician clinicians in the United States. N Engl J Med 348:130-137

Groopman J (2000) Second opinions: Stories of intuition and choice in the changing world of medicine. Viking, New York

Glouberman S, Zimmerman B (2002) Complicated and complex systems: what would successful reform of Medicare look like? Commission on the future of health care in Canada, Discussion paper No. 8, July 2002

Kleinman A, Eisenberg L, Good B (1978) Culture, illness and care: clinical lessons from anthropologic and cross-cultural research. Ann Intern Med 88(2):251-258

Medical education issues of (1981-2002) JAMA

Nelson EC, Batalden PB (2000) Chapter 4. Knowledge for improvement: improving quality in the micro-systems of care. In: Goldfield N, Nash DB, (eds) Managing quality of care in a cost-focused environment. ACPE/Aspen Publication, Frederick, MD

Quinn JB (1992) Intelligent enterprise. The Free Press, New York

Rothstein WG (1987) American medical schools and the practice of medicine: A history. Oxford University Press, New York

Schön DA (1983) The reflective practitioner: How professionals think in action. Basic Books, New York

Schön DA (1987) Educating the reflective practitioner. Jossey-Bass, San Francisco

Schein EH (1992) Organizational culture and leadership, 2nd edn. Jossey-Bass, San Francisco

Shaw P (2002) Changing conversations in organizations: A complexity approach to change. Routledge, London, New York

Weick KE (1995) Sensemaking in organizations. Sage, Thousand Oaks

Weick KE (2001) Making sense of the organization. Blackwell, Oxford

Weick K, Sutcliffe KM (2001) Managing the unexpected: Assuring high performance in an age of complexity. Jossey-Bass, San Francisco

Weick K, Sutcliffe KM, Obstfeld D (1999) Organizing for high reliability: processes of collective mindfulness. Research in Organizational Behavior 21:81-123

14 Organization and Leadership in Hospitals

James H. Taylor[1]

Good Morning. I was surprised yesterday morning after listening to Karl Weick's wonderful presentation to find myself thinking that I should be telling a story, rather than making an attempt to give an academic lecture. I am not an academic and most of the communicating I do is, in essence, story telling. Listening to Dr. Weick, I was struck by how effective his narrative was in making his points. If an outstanding scholar can stand before so many accomplished academics and tell a story, I decided that it would be all right for this practitioner to do so also. So I left the disk with my PowerPoint presentation in my room. I hope you will find my story interesting.

I have been a teaching hospital CEO for about twenty years. I am not a clinician. I don't deliver care. My interest is in organizations and leadership. I am classically trained, by that I mean I have an MBA and an MHA; I have worked for a steel company; and I spent four years in the military during the Viet Nam era. I have every reason to think about organizations and leadership in what I call a classical way. Eight or nine years ago I hooked up with Curt Lindberg and some of his colleagues, and I began to think about different ways of understanding what was going on in my life in organizations. Until that time, there was a big gap between what I had been taught should be happening and what I was experiencing. During the last eight years, I have had the chance to look into complexity; and over the last three and one-half years, I have thought and read and written about complexity and organizations as a doctoral student.

What I would like to do in the next few minutes is tell a story about unexpectedness and surprise that happened in our organization a couple of years ago and then to reflect on my part in this story in several ways: One way, the way that a board chairman would likely reflect on my performance and as would most of my colleagues who have been trained as I have been. And then, from a perspective that is complexity informed. And then having done that, pose the question, "If you make sense of something in a different way than other people, so what?" Then finally, suggest a question about research that occurs to me from both this experience and the last few years of my educational activities.

So, let me tell you my story. I work at the academic medical center in Louisville. We are an inner-city teaching hospital. A couple of years ago, the day before New Year's Eve day about lunch time, walking by the fax machine, I noticed a document from the federal government agency that regulates hospitals and the Medicare system. Picking it up, I assumed it was just a regular, yearly update on the regulations they had issued and perhaps a preview of the ones that would come next year. As I read it, a phrase caught my attention. This fax said the federal government had determined that the University of Louisville Hospital presented an

[1] James H. Taylor, President and CEO, University of Louisville Hospital, jtaylor@ulh.org

immediate and serious threat to patient health and safety. And having determined that, they were going to publish in the local newspaper a notice to the public saying so. Further, upon publication of the notice that we were an immediate and serious threat to patient safety, the government would cease paying for any services rendered to Medicare patients.

The week between Christmas and New Year's is usually a rather slow time. Mostly, nobody is around. Now all of a sudden, we had one and one half working days to deal with this issue. The fed's notice indicated we could appeal this action if we could provide a credible allegation of compliance, whatever that meant. If we were successful, they would then publish in the newspaper, within twenty-one days, the fact that we were okay again; and people on the Medicare program could come back to University Hospital (and we would get paid to take care of them). It appeared that we were in big trouble, and our standing in the community was about to be irreparably harmed.

I walked down the hall to my colleague, Kay's, office who fortunately was one of the few who was in the office and happened to be the individual whose responsibilities included the issue before us. We talked briefly. She immediately took responsibility for trying to gather nursing and business people to see what could be done to understand the issue. I called the hospital attorney, mostly to find out what a credible allegation of compliance really meant. He said that he had had some previous dealings with the attorney in the fed's regional office in Atlanta. He suggested he call his colleague and explore what options we might have. He did so, and about an hour later, he called back to say, "There is some good news in that you can avoid this sanction if, within twenty-four hours, you can make your argument of compliance (in the governmental way, of course, on their forms) to Atlanta. And if you are successful, they will not publish that you are a threat to patients." That sounded like a very good outcome.

We immediately informed the gathering group, now of about eight people, that we had about six working hours to convince the feds we were not a threat to patient safety. The initial research had by then revealed that the source of this problem was a single psychiatric patient who had left our hospital voluntarily, but without signing out six weeks earlier. We had reported the incident, the state had investigated, and we thought the matter had been resolved. The state was required to submit this information to the feds and they had done so. Without any further consultation with us or the state agency, the feds had made their determination that we were a serious and immediate threat to patient safety and had initiated action to inform the community. I have since wondered more than once what would have happened if I hadn't walked by the fax machine that afternoon.

The work facing us was clinical and rather technical. I was not taking part in the interactions going on in the workroom because I had nothing to contribute to the substance of the work that needed to be done. Nevertheless, I was very interested in what was going on in that room. I was trying to be present without being in the way or appearing to be looking over people's shoulders. There was a big buzz in the workroom, lots of conversation, and lots of back and forth about who was going to do what and what needed to be done.

The group worked for about three and one-half hours that day and then left for the evening. They agreed to come back on New Year's Eve Day; many of us had other plans, but everyone would come back. Kay, the convener of the group, was rather nervous as they all left. She had yielded to the group's sense that they could get the work done the next day in time to meet the noon deadline, but she obviously was very uncomfortable with the decision. The next morning, she and I were there by 7:15. The rest of the group moseyed in, and the work finally was underway by 8:30. I sensed a growing tension in the group as the morning progressed.

By 11:15, forty-five minutes before the deadline, the word processing people were humming. They were dealing with pieces of paper coming in and out of the room, trying to understand what was being said, and frantically trying to find the right places on the forms to put the text. Around 11:45 as I was wandering around, trying to sense how things were going, I observed Kay at work. I noticed her neck was flushed, the tone of her voice was close to quivering, and the way she was dealing with people was definitely not characteristic of her normal thoughtful and calm demeanor. It was clear to me that we weren't going to meet the noon deadline. And it was also clear to me that she wasn't going to tell me that.

We had a brief conversation during which I suggested that she call the federal official and ask if we could present our information in our own format, since the problem we were having was not with content or substance, but rather in trying to comply with the governmental formatting of the information. She headed for her office to make the call and returned rather quickly to report that the feds would give us an hour to get the information faxed to them in our format. We would be allowed to submit the information in their format by overnight mail.

The information was faxed at 1:00 p.m. The work group went off to have their holiday. Kay and I sat around waiting for the phone call from Atlanta. The call came about one hour and forty-five minutes later. The federal official who called us said, "We are not sure whether you are appealing our action or whether you are accepting what we have done and making a case that there isn't a problem anymore. If you are appealing, it is not within my jurisdiction to make a decision. You will have to wait three weeks for resolution and the notice will get published as now scheduled. However, if you will accept that the action we have initiated was proper, I can make a ruling as to whether your local responses have resolved the issue and, if so, withdraw the action, which means the notice would not be published."

We got our attorney back on the phone, and we had a three-way conversation. I assured the federal official that we were not appealing; our attorney told us that we could do that later if we needed to. But the important thing was to stop the notice from being published in the newspaper. We said so, and the federal official said she would be back to us within the hour with her decision. About ten minutes before 4:00 p.m., she called to tell us she had ruled we were in compliance and the notice would not be published. We thanked her for her attention to our issue and wished her a Happy New Year.

As Kay and I and the administrative assistants working on the governmental forms finished up and prepared to leave, the hospital's attorney arrived with two bottles of champagne. I made an immediate executive decision that we would temporarily suspend our self-imposed rule that there would be no alcohol consumed in the executive offices. We polished off the two bottles of champagne pretty quickly amidst toasts to our good fortune, marked more by relief than any feeling of accomplishment. We departed together, all of us intent on having a good New Year's Day.

So that is my real life story of surprise. It wasn't a very pleasant surprise, although things turned out okay in this case. What I have done since this experience is to reflect on what had been going on, trying to make sense of both the situation and my actions as CEO.

I have thought about how a board chairman, who is typically a successful, classically trained male business leader, would likely make sense of this experience. I suspect that he would have seen the surprise of the fed's threatened intervention as a failure of our organization to anticipate a problem, or at least failure to have information systems in place that would have warned us of a problem so we could suppress it or at least remedy the default before we were sanctioned. This perspective is informed by the classic view of the organization as a machine whose actions can be predicted, planned, and controlled if the organization is properly designed, trained, and managed.

I think a board chair would question from a classical theory of management and organization whether I, as CEO, responded appropriately. I was not present in the room where the work was being done, I wasn't giving people orders, I wasn't closely supervising the work process at a time of potential organizational disaster. I didn't get in that workroom and say, "I want some answers now - what happened and what are we going to do about it so as to avert the fed's sanction?" I didn't take charge. I think that a typical board chair would have expected that of me.

Perhaps my actions could be explained by the viewpoint that I was taking a "big picture" perspective. That is what CEOs are supposed to do, right? Our hospital was going to be in serious trouble and so maybe I didn't need to be involved in the detail work at all. After all, the workers were doing the work and there was another senior manager to supervise them. The CEO should be paying attention to what was going to happen to the organization's reputation and public standing. And, indeed, I really did do just that, spending a lot of time during that day and a half with our public relations director, getting ready to explain to the public what the newspaper notice meant.

The truth is that I was paying attention to both the work going on in the workroom and, at the same time, I was concerned about the "big picture" of public relations. I just didn't go about doing those things in what a board chair might consider the conventional way. I was aware that what was going on in that workroom was really messy. There were a lot of people who weren't easily agreeing with each other; there was conflict. A board chair could have expected me to resolve that conflict, to get them aligned so as to get the work done. I didn't do that; I didn't exercise my formal power to take control of the situation. Was I a negligent or irresponsible leader? From a classic perspective, a board chair might well judge my actions as such and conclude that we were just lucky to get out of this situation unharmed.

Let me move now to how I have thought about this experience. While I am not going to explicitly cite complexity theory, I think you'll see how complexity ideas have informed my sense making. As I was living it, I never considered organizational failure as the cause of the situation; rather I approached the time as an unexpected and unpredictable emergent event that needed to be dealt with. I saw my job as engaging with those involved in whatever way the changing circumstances might suggest to me. My focus was in the present of that time, that is to say my attention was directed to what was occurring in the moment around me. I was not focused on what should have happened, or on what might happen, but rather what was happening with those with whom I was involved. That is not to say that in the moment of that time that the conversation didn't consider the past or present, only that my attention was sharply focused on what those involved, including myself, were saying and doing. I understand the past to be only known in the present and the future to be created in the present. Thus, my focus was on the here and now of the moment. Karl Weick talks about such attention to the moment as "mindfulness."

Believing the conflict that was going on in the workroom as necessary for anything useful to emerge from a group of people not used to working together, I did not step in to try to resolve the differences. I saw the differences as diversity of perspectives and opinions and believe it is from such diversity that novelty can arise. And at that point, we needed novel approaches to the situation we found ourselves in. None of us had been there before.

So what difference does it make what way I make sense of what happened in the surprise I have described? After all, everything turned out okay. For me, how I make sense of what happens to me will inform the way I will participate going forward. Sense making informs action and how I act will inform the people I work with about how I can be expected to participate and, in turn, how they will make sense of our interactions together. But my sense making is also informed by how I understand others' actions. Sense making, for me, is a socially constructed process. My sense making informs others and is, at the same time, informed by others. I think that's a different way of seeing my job - as someone who contributed to the sense making of that time, rather than someone who knew what to do and who got it done. Knowing what to do and getting it done is what CEOs are all about, if one buys the classic way of thinking. The reality is that I didn't know what to do. The sanction would not have been averted without that group of people figuring out together what needed to be done and getting it done. That, I think, amounts to an ability to make individual and collective sense of a situation and find ways to go on together in the work at hand.

Let me end with these questions: What might this way of relating my experience and my way of understanding that experience offer to how we think about research in the social sciences and in human organizations? Is narrative accounting of experience and the process of sensemaking of that experience a form of research? What is the test of valid scientific inquiry? For me, it is found in my ability to make sense of some physical phenomenon or personal experience. In the telling of my experience, has it contributed to how you think about your own experience or, put in my way of saying it, to your own sensemaking? If it has, is this reflective narrative not then a valid form of research?

The classical way of thinking about research validity is in its replicability. If you believe, as I do, that human experience is not replicable in the way that a classic scientific experiment is, what is the measure of valid research of human activity, like organizations? I am suggesting that the test is sensemaking of the experience and that the narrative of one person's experience and sensemaking of it can contribute to others' processes of sensemaking and, therefore by my definition of the validity of research, the kind of reflective narrative I have presented can be an important form of social research.

Thank you for listening to my story.

IIID Fundamental Uncertainty in Business and Business Decision Making

15 The Fundamental Uncertainty of Business: Real Options

James S. Dyer[1]

The purpose of this paper is to discuss the manner in which uncertainty is currently evaluated in business, with an emphasis on economic measures. In recent years, the accepted approach for the valuation of capital investment decisions has become one based on the theory of real options. From the standpoint of this workshop, the interesting aspect of real options is its focus on the flexibility of management to respond to changes in the environment as a feature of an alternative that has unique value, known as "option value." While this may not be surprising to most participants in this workshop, it does represent a radical change in traditional thinking about risk in business, where efforts have primarily been focused on the elimination of risk when possible.

Prior to my presentation, Henri Lipmanowicz posed a question and requested a response, so we begin this discussion with that exchange. Next, real options are introduced and discussed at an elementary level, but with an emphasis on the value of managerial flexibility as a response to uncertainty and the possibility of a surprise.

15.1 The Question

The question from Henri Lipmanowicz that I addressed might be paraphrased as follows: "How do you distinguish between the bad surprises (namely the performance problems) that arrive from fundamental uncertainty, meaning they are unavoidable; versus those that are the consequences of the organization and the processes that you have put in place? In the latter case, something could have been done to avoid the bad surprise, or we may be able to learn how to avoid similar bad surprises in the future, but that may not be true in the former case."

This question reminded me of an experience from several years ago. At that time, I was sitting in my office, quite happy, thinking about things that scholars think about, and that may not be all that relevant in the real world. I was surprised by a telephone call from a lawyer, and I suspected that it might lead to an interesting experience. The telephone call went something like this. The lawyer said, "Jim, I was given your name as someone who could be an expert witness for us before the Public Utility Commission of the State of Texas." I expressed some interest, but asked for more information. He said, "Have you ever heard of the South Texas Nuclear Power Project?"

[1] The Fondren Foundation Centennial Chair in Business, McCombs School of Business, The University of Texas at Austin, j.dyer@mail.utexas.edu.

The South Texas Nuclear Power Project is one of two nuclear power plants that were built in the State of Texas, and both projects were started in the late 1970's when natural gas prices were relatively high. They were both subject to significant cost overruns, and were very controversial at the time, with several efforts being made to cancel them prior to their completion. Many of these issues had been documented in local and national newspapers, and the basic story was well known.

The lawyer continued, "Well, my client is an electric power utility company, and we have to go before the Public Utility Commission of Texas to participate in a prudency hearing regarding the expenses incurred in constructing the South Texas Project." The Public Utility Commission allows utility companies to construct power plants, and to increase rates sufficiently to cover the cost of constructing these power plants. However, when the plant is completed and actually begins producing power, then the Commission holds a hearing to decide whether all the costs that were incurred in the construction of that power plant were "reasonable and prudent," then they are "disallowed" and the utility company is not allowed to recover them from the rate payers.

And the lawyer said, "We have to go before the Public Utility Commission and convince them that all the decisions we made in the context of a very risky and somewhat chaotic environment associated with nuclear power in the 1970's and 1980's were reasonable and prudent. And we want you to provide testimony in support of that effort."

I asked for some additional information, and he said, "Well, here is the situation. When we started the project, we estimated the plant was going to cost $800 million." And I asked, "So, what was the final total?" He responded that the final cost of the plant was somewhere between $6 and $7 billion. I summarized that point, by saying, "You want me to go before the Public Utility Commission and tell them that your client started out with a plant that was expected to cost $800 million, but it ended up costing between $6 and $7 billion; and that all the decisions that were made were reasonable and prudent." And he responded, "That's the idea." I took a deep breath, and said "Well, I will do that. But I can tell you that the effort will be a costly one."

The effort to demonstrate that decisions were reasonable and prudent at the time they were made requires an extensive effort to review those decisions, based on the information that was available to the decision-makers when those decisions were made. The key question is the following one. Would other, reasonable people have made different decisions when faced with the same information that those decision-makers had at the time the decisions were made? It is a challenging task, in some cases, to recreate history, particularly when people do not leave detailed documentation of the decision processes they used and the information that they did have. They may not provide a discussion of how they actually rationalized or made sense of the decisions they faced.

In more recent years, at least in the electric power industry, employees have gotten much better about documenting decisions; documenting the information they had, documenting the process of decision making, and being sure that a record is available so that someone can look back, and try to make sense of the way the decision was made. When this record is available, then one can make a comment about whether a decision was a reasonable and prudent one in the face of risk and uncertainty, even if it turned out that the decision resulted in a bad outcome, a bad surprise.

Further, it is very important to document the basis for these decisions as a part of the decision process, at the time the decisions are made. Otherwise, these decisions will be reviewed by individuals looking back at decisions with the advantage of hindsight. When a decision results in a bad outcome, it often seems clear in hindsight that the decision process was flawed and imprudent, as documented by the vast literature on the "hindsight bias." So the process of documenting the rationale behind decisions in the face of risk and uncertainty is a very important one, and this concept is emphasized in the evaluation of these decisions at the Public Utility Commission.

15.2 Real Options

The notion of documenting a decision that is reasonable and prudent implies some standard for an appropriate analysis of a decision. Many investment decisions are made in the context of risk and uncertainty, and the accepted methods for valuing alternatives in this context has changed over time. In particular, the approaches to dealing with uncertainty and surprise have matured a great deal in the past two decades.

For example, how would a company be expected to evaluate the following question: Should we build a nuclear power plant, given that there is uncertainty in the demand for power, the costs of nuclear and competitive fuels, the construction costs, the regulatory environment, and the concerns regarding environmental hazards? This is a very complex question, of course, so it is instructive to start with a simpler question: How should one estimate the value of a business investment? The starting point for answering this question is a forecast of the cash flows that will be generated by that investment decision over time. Next, the appropriate discount rate is applied to that cash flow stream to obtain its present value. For example, the discount rate might be chosen to be ten percent. The cost of the investment is subtracted from the present value of the cash flow stream to obtain the "net present value." This approach to valuing a project is straightforward to implement, and it is called the discounted cash flow (DCF) method. This discounted cash flow, or net present value, is taken as an estimate of the value of the project, and this approach was used in most of the studies regarding the estimated value of completing the South Texas Nuclear Project in the 1970's and 1980's.

This approach works reasonably well for some projects, particularly when there is no risk associated with the forecast of the cash flows and no anticipation of surprise. On the other hand, when there are risks associated with the estimates of the cash flows, and even the possibility of a surprise, then the problem of estimating the value of an alternative becomes a bit more difficult.

These complexities are addressed today by using concepts associated with a field called "real options." It is important to understand the concept of an option.

Figure 15.1 shows a risky investment decision illustrated by using a decision tree. The decision tree is used in business investment decisions to model where decisions are made, and also the risks that might be faced. In the simple decision tree in Figure 15.1, a square represents a decision, or a choice between two alternatives. For example, we may have the opportunity to invest in a nuclear power plant, or not to invest in a nuclear power plant. In Figure 15.1, the decision is made first, and then we wait and to see what happens regarding some risk.

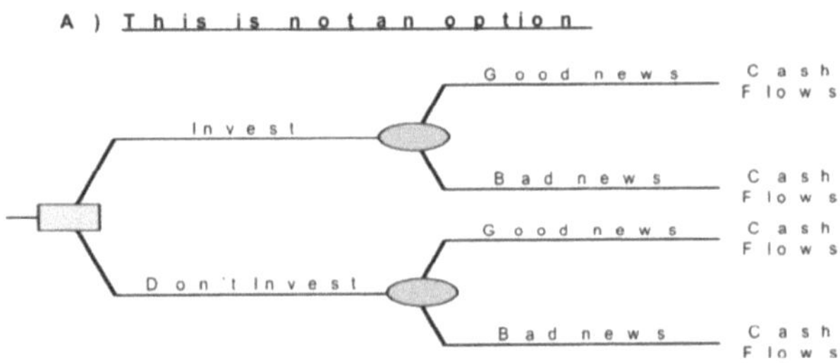

Fig. 15.1. A decision tree representation of a risky decision

If there is some risk, then the news might be good; and so our cash flows might be high. Or the news might be bad, and the cash flows might be low. So when the decision involves some risk, the cash flows cannot be predicted with certainty. Nevertheless, this decision tree indicates that we have a choice; we can invest, or we do not invest. After the decision is made, the risk will be resolved and the outcome may be good or bad.

In the 1970's, a risky alternative such as the one shown in Figure 15.1 would be evaluated by taking the expected values of the risky cash flows. For example, there might be an estimate that the probability of good news and high cash flows is 0.5, and the probability of bad news and low cash flows is also 0.5. The expected value would be computed by multiplying the cash flows in the good case and the cash flows in the bad case by 0.5, and by adding these probability-weighted cash flows to obtain the expected value. That is the way risky projects were typically evaluated at that time.

For some projects, that approach is not an adequate representation of the complexity associated with the problem and of the options that may be available. This traditional approach implies that, once a decision is made to invest, a manager cannot react to changes in the environment. That is, the assumption is implicitly made that managers have no flexibility associated with how the project is actually managed over time. This is an unrealistic assumption for many projects, and therefore leads to errors in estimates of their economic value.

In contrast, Figure 15.2 provides a simple diagram of a project with options. Now this is the way a manager would like to make investment decisions. That is, first the manager would find out whether the news will be good or bad, and then she would make a decision. In this situation, the decision maker has an option. She can exercise the option to either invest or not after knowing whether good news (with high cash flows) or bad news (with low cash flows) has occurred. And options have value.

This new approach to the valuation of projects is called "real options" because it is related to the notion of how one prices a stock option. The theory and methodology for pricing stock options is based on work done over twenty years ago by Black and Scholes. For example, you can either buy a share of Dell stock or you can buy an option on that stock. A call option on Dell stock could give you the right to buy 100 shares of the stock for $30 per share at any time over the next three months. And then, if the price of the stock goes above $30, you have the right to exercise the option, and take your profit. If the price of the stock does not go above $30, then, of course, you lose the money that you paid for the option. What would a stock option like this be worth?

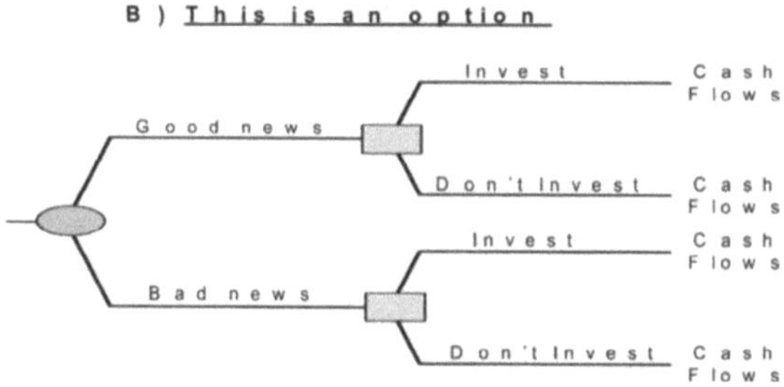

Fig. 15.2. An investment decision that is a real option

If we do have an option to make a decision after we know what has happened, then the investment decision is much easier. Therefore, the value of being able to make a decision after a risk has been resolved, as illustrated in Figure 15.2, is higher than in the previous situation illustrated in Figure 15.1, where we have to decide, and then wait and see what happens. Therefore, it should be clear that business decisions including options to make decisions after risks have been resolved should be worth a premium, since these options add to the value of investment decisions. However, this adjustment for the value of flexibility in business decisions has only been explicitly calculated in recent years.

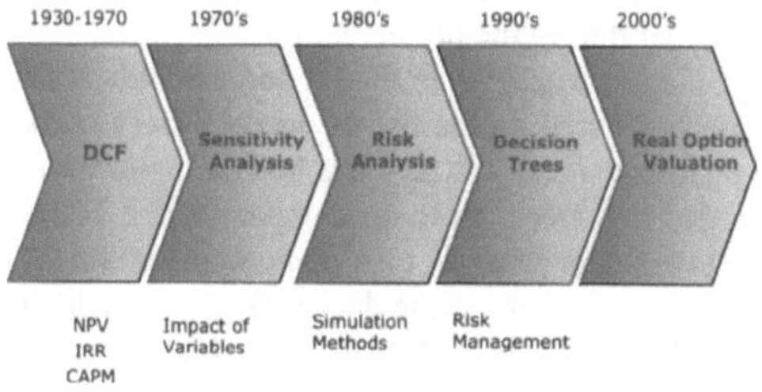

Fig. 15.3. Evolution of approaches to the evaluation of business investments

Figure 15.3 provides a simple diagram of the progression in approaches for valuing business investment decisions. Discounted cash flow (DCF) became popular in the 1930's. As noted earlier, DCF is sometimes called net present value, and is also related to approaches for estimating the internal rate of return (IRR) for a project. The evaluation of alternatives based on DCF requires arguments based on the Capital Asset Pricing Model (CAPM) which is a cornerstone of modern financial theory.

The availability of computers in the 1970's allowed analysts to begin doing some sensitivity analysis associated with their estimates of project values. A sensitivity analysis begins with a computer model of a business investment decision that estimates cash flows and the DCF, but then allows the user to change some parameters and see if the DCF also changes significantly. That was the advanced approach to dealing with uncertainty in investment decisions at that time. What if the interest rate goes up? What if the revenues are not as high as we anticipated? What if another competitor comes into the market? What if? That is the way that an analyst dealt with uncertainty in the 1970's, given the availability of mainframe computers.

In the 1980's, people began to do risk analysis, using Monte Carlo simulation. Software programs became available that would allow the user to insert probability distributions into spreadsheets, and to run these models on laptops so that uncertainty about spreadsheet parameters could be incorporated. This was an improvement over a simple "what if?" sensitivity analysis, since the ability to introduce probability distributions in spreadsheets allowed the user to obtain probability distributions over net present values. This allowed managers to consider both the risk and the return associated with investment opportunities, and to make the tradeoffs between these two concerns.

But in all three of these methodologies, there was very little recognition of flexibility in projects based on decisions that management might make after events have occurred. And this flexibility associated with projects was not a part of the analysis, and so the value offered by the option to abandon the project, or to expand it, was not included in the estimate of the project value. However, in the 1990's, the software became available to allow analysts to consider flexibility, and to model flexibility using sophisticated decision tree models. This approach has been adopted in many industries, because decision trees allow the user to recognize decisions that can be made later, contingent on uncertain events that either occur or do not occur during the project's life.

In particular, many companies in the oil and gas industry have adopted project valuation approaches based on the use of decision trees to model project flexibility. Also, the pharmaceutical industry has been a major user of risk management tools based on decision trees. How do companies like Merck, Bayer, and Pfizer decide whether to conduct research on a compound that might lead to a new drug? How do these companies make those decisions given the high costs of identifying a compound that has promise in the lab, the costs and risks associated with the phases that are required to obtain FDA approval for a new drug, and the market risks associated with uncertain demand? How do companies screen the thousands of possibilities to identify those particular drugs that are the most promising investments? Decision trees play a major role in the evaluation of these potential new products.

As we discussed earlier, real options provide a related approach to evaluating project flexibility. Project flexibility may take the form of an option to expand the project, or to abandon it, and these decisions can be viewed as analogous to decisions to exercise different forms of stock options. Further, this analogy provides some interesting insights that may not be intuitively obvious.

Suppose you do have options in a business project. What are the values of the options associated with a project? The value of an option depends upon two things, as shown in Figure 15.4. First, it depends upon the level of flexibility that the option provides. Flexibility associated with decisions adds value to real assets. Second, and perhaps somewhat surprising, is the insight that options are more valuable when they are associated with an investment decision with a high degree of uncertainty. This latter point is somewhat counter-intuitive, because the objective of many actions taken by managers is to reduce business uncertainty. But if you have flexibility, including the option to abandon the project, for example, then it may be better to invest in very risky, uncertain alternatives that have the potential for surprise. This is because you can reap the rewards of the positive surprises, and drop the project in case of negative surprises. So that insight is an important one, and it is counter to the usual business objective of eliminating uncertainty.

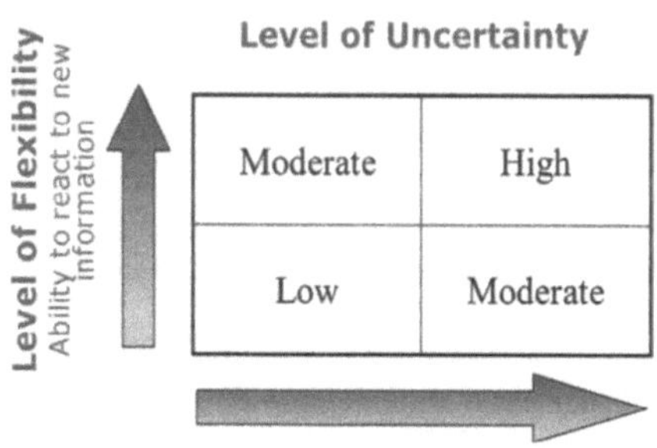

Fig. 15.4. Conditions under which the value of a real option increases

15.3 With Flexibility and Uncertainty

There are two issues associated with the application of real options concepts to business projects. One of them is the need to model project flexibility, and the other is the methodology for valuing the risky cash flow streams.

15.4 Modeling Project Flexibility

Modeling flexibility is a task that can be supported through the use of decision trees. Figure 15.5 illustrates a simple decision tree for an offshore oil and gas development project. This example is an easy one to understand, since the oil and gas business is very risky, and some of the risks are easy to identify. If a company drills a well, it might be a dry hole. Even if the well is successful in finding oil or gas, the size of the oil reserves may be very uncertain.

Figure 15.5 only shows two uncertainties. One is the oil price, and the other is the size of the oil reserves, typically measured in barrels. However, there is no flexibility shown in Figure 15.5. It would appear that a manager would simply make a decision to drill the well or not, and then would passively wait to see what occurs regarding the actual reserves that are found and the price of oil that occurs.

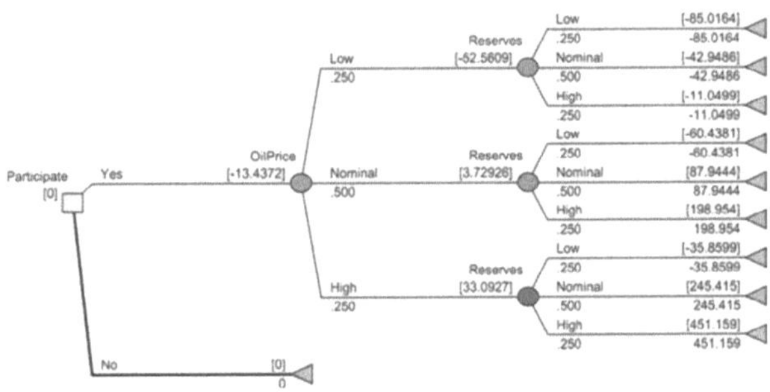

Fig. 15.5. A simple decision tree with no options for an oil and gas exploration project

In contrast, Figure 15.6 shows a decision tree that does include a number of management options, and these options provide flexibility for the project. This is the way that projects are modeled in order to include the value of options, or project flexibility, in the valuation of projects. The details of the decision tree are very difficult to read, but the small squares are decisions that can be made over the life of the project, and all but the first two squares represent the decisions that can be made after something is learned. These decisions represent the flexibility to change the way the project is managed, or even to abandon the project, on the basis of new information.

An example of an option may be stated as follows, "Should we drill the well now, or should we wait until oil prices change?" Another option to consider would be, "If we go ahead and drill now and the first well is not successful, should we drill a second well?" Or, "If the project turns out to be successful, and oil prices change, when do we terminate the project?" Building this flexibility into the evaluation of a project is important, and it does change the value of the project, and can also be used to determine the decisions that should be made as the project is managed over time.

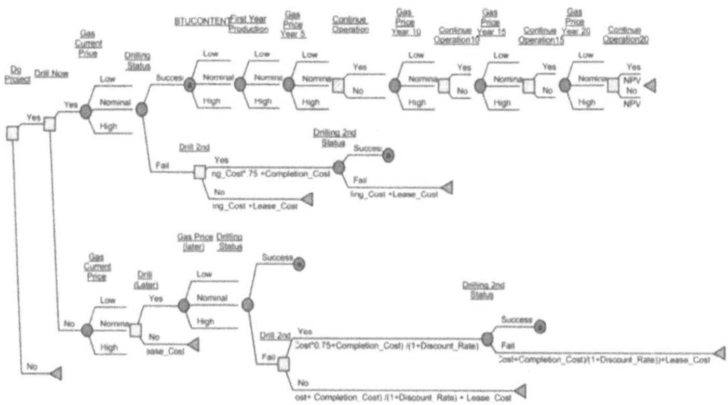

Fig. 15.6. A decision tree model of an oil and gas project

15.5 With Options

Recognizing and modeling flexibility is very important, and there are techniques based on decision trees and related ideas that provide a means of creating the appropriate models. Further, options in a project are really very difficult to value in an intuitive manner, and so these models are necessary in order to assess the true value of projects that offer flexibility to management. This flexibility typically changes our estimates of the net present values of projects in a very positive way, because the manager can eliminate downside risk if the option to abandon the project is available, for example. So projects with options are desirable.

15.6 Valuing Risky Cash Flows

Finally, it may be interesting to consider how risk itself is valued in business investment decisions. The simple decision tree in Figure 15.5 also illustrates some important distinctions regarding business risks. For example, if drilling a well is successful and oil is found, then the uncertainty regarding the reserves is reduced over time, and essentially disappears because one learns what the actual reserves are. However, the oil price risk is always there. The price of oil goes up and down over time, and that risk cannot be eliminated during the lifetime of the project. So some risks may disappear over time because of learning; other risks stay with you for the entire life of the project.

These two risks are also very different in another way. According to the theory of finance, stock market investors should only care about one of these two risks. That is, there is only one of these two risks that would cause an investor to pay less than the expected value of this project because the risk is to be avoided. Which one would bother an investor? In theory, an investor should not care about the risk of reserves associated with an oil and gas project. The only thing the investor cares about is the price of oil, which may seem like a surprising idea.

The rationale for this notion is the following one. When investment decisions are considered from the standpoint of stockholders, we assume that the rational stockholder will be diversified. Therefore, a rational stockholder will own stock in Exxon, Mobil, Chevron, Texaco, and so on, and each of those companies drill hundreds of oil wells. Because so many wells are drilled, the investor should obtain his or her share of the sum of the expected values of each of these wells. The actual reserves on one well may be higher than its expected value, and the reserves on another may be lower, but the sum of the actual reserves should be close to the sum of the expected values of reserves for all of the wells. So the risk associated with the reserves on any one well does not matter, because that risk gets averaged out, as long as the risks associated with the sizes of the reserves associated with different wells are probabilistically independent.

On the other hand, the oil price risk does not "average out," because any changes in the oil price will affect all of the wells in the same way. And furthermore, it turns out that the price of oil is correlated with the stock market, and that creates problems for the portfolio of the investor, since that increases the risk of the portfolio.

So one of the interesting things about investment decisions is that some risks matter to the investor, and some risks do not matter. And these types of risks are treated differently in terms of the way projects are valued.

The traditional approach to adjusting the value of a project to reflect the risk associated with it is to estimate the expected cash flows that it will generate over time, and then to discount those cash flows using a risk-adjusted discount rate that is related to the correlation of the risks of the project with the stock market. This risk-adjusted discount rate is determined by finding another security traded in the market, typically the stock of some company, that has essentially the same risk characteristics as the project. Information based on the market price of the stock can be used to estimate the appropriate risk-adjusted discount rate for the stock and, therefore, for the project with similar risks.

A naïve approach to valuing projects with real options would be simply to include decision nodes corresponding to project options into a decision tree model of the project uncertainties as shown in Figure 15.6, and to solve the problem using the same risk-adjusted discount rate appropriate for the project without options. Unfortunately, this naïve approach is incorrect because the optimization that occurs at the decision nodes changes the expected future cash flows, and thus, the risk characteristics of the project. As a consequence, the standard deviation of the project cash flows with flexibility is not the same as that of the project without flexibility, and the risk-adjusted discount rate initially determined for the project without options will not be the same for the project with real options. This fact has caused some authors to wrongly conclude that is inappropriate to use decision trees to value real option problems.

However, decision trees may be used with risk neutral probabilities to solve real option problems. This implies that we can discount the project cash flows at the risk free rate of return and make any necessary adjustments for risk in the probabilities of each outcome.

An example illustrates this concept. Suppose the risk free rate is 8%, and that there is a simple project with equal chances of cash flows of $170 or $65 one year from now that has a risk-adjusted discount rate of 17.5% and that will cost $115 next year. For obvious reasons, the events associated with these two outcomes are commonly called the "up state" and the "down state," respectively. These two states have each been assigned objective probabilities of 0.5. The expected present value of the project is [0.5($170) + 0.5($65)] / 1.175 = $100 and the net present value is -$6.48 as shown in Figure 15.7.

Suppose now that the decision to commit to the project can be deferred until next year, after the true state of nature is revealed. The original discount rate of 17.5% cannot be used because the risk of the project has now changed due to the option to defer the investment decision. On the other hand, a set of risk neutral probabilities for the original project (probabilities that would give the same project value as before when discounting the cash flows at the risk free rate of return) can be determined and used to value the project with the deferral option, since the expected cash flows for both problems are the same ($170 and $65).

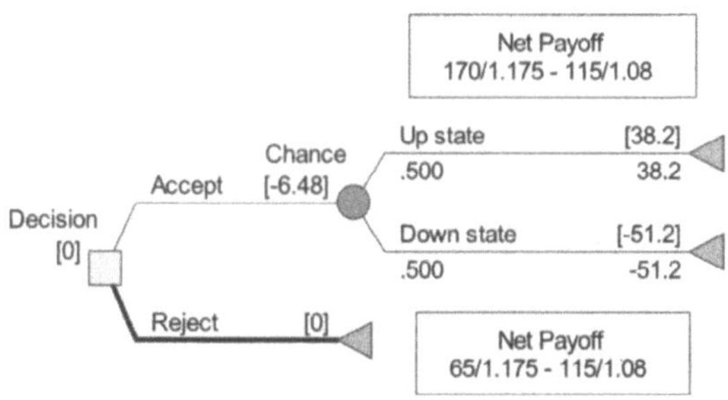

Fig. 15.7. The project with objective probabilities and a risk-adjusted discount rate

While the correct risk-adjusted discount rate of a project with options is difficult to determine due to the effect these options have on the project risk, the risk free rate of return can be readily observed in the market. By switching from objective probabilities to risk neutral probabilities, the project net present value with options can then be estimated even without knowing the correct risk-adjusted discount rate.

In this simple example this can be done by setting the expected present value of the project determined with the objective probabilities and the risk-adjusted discount rate equal to the expected present value of the project with the unknown risk neutral probabilities and the risk free discount rate, and by solving for the risk neutral probability p. That is, we would let

$$\$100 = \frac{p(\$170) + (1-p)(\$65)}{1+0.08}$$

and solve to determine $p = 0.41$. In practice, these risk neutral probabilities are obtained from data available in the market.

The project with the option to defer has net payoffs of $170 - $115 = $55 in the up state and zero in the down state as illustrated in Figure 15.8, as there will be no investment if it is known beforehand that the down state will prevail. The net present value of the project with the option to defer is [0.41($55) + 0.59($0)] / 1.08 = $20.86, up from -$6.48. This implies that the value of the option to defer is $27.34.

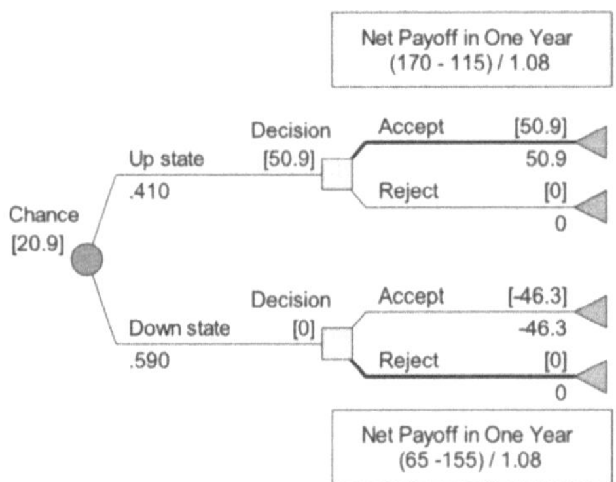

Fig. 15.8. Project with risk neutral probabilities and a risk free discount rate

15.7 Conclusions

The evaluation of risky investment decisions is based on methodologies that have evolved over time to adjust for the risk of a project, and to adjust for flexibility as well. The initial advances in project valuation were based on using the appropriate discount rate to determine a net present value for the project. More recently, the tools of decision trees and real options have allowed managers to determine the appropriate adjustments in estimates of project values that reflect flexibility, or the opportunity to react to uncertainty and surprise. This flexibility has value, and these methods provide a sophisticated approach to estimating its worth.

For more information regarding these ideas, please refer to the web site: http://www.mccombs.utexas.edu/faculty/luiz.brandao.

16 Transforming Your Regional Economy through Uncertainty and Surprise: Learning from Complexity Science, Network Theory and the Field

June Holley[1]

16.1 Introduction

The field of regional development blossomed in the last decade, as researchers and practitioners increasingly asserted that the region was the most effective geographic unit for supporting the excellence and innovation of entrepreneurs.[2] Much of the discussion of regionalism continued to be mired in concepts and language of the industrial age. Many regions started their regional renewal processes with large convening of area power brokers, who create a common vision of the future of the region and then develop a plan intended to move the region toward that vision. Uncertainty is reduced as much as possible and surprise is viewed as unwelcome. Unfortunately, this type of linear, rational process is seldom effective in dealing with uncertainty and an unknown future, and has had little success in solving the massive problems of poverty and environmental degradation that continue to plague inner cities and rural communities.

This paper suggests that complexity science, the study of the dynamics of complex systems, may provide a more useful framework to guide change processes. At the same time, complexity science can now offer insights into the qualities underlying a healthy resilient economy, and the processes by which a rapid shift to such an economy might be catalyzed.

16.2 Images of How Regional Change Happens

Examples of transformation in healthy, resilient, natural systems can be used as metaphors to help one to think creatively about the process by which healthy regional economies emerge. To transform regional economies, small collaborations

[1] President/CEO, Appalachian Center for Economic Networks (ACEnet), juneh@acenetworks.org.

[2] See, for example, the many studies by the European Union and the work by Michael Porter.

among a few people can co-create the beginnings of a new economic ecosystem, attracting entrepreneurs and leading to a cascade of innovation.

A vivid example of transformation that shows the power of self-organization is the construction of termite castles. Termite castles start with the simple act of digging tunnels. Little mounds of dirt begin to accumulate, and groups of termites simply start building on those mounds. If what they do keeps them cool and moist (the large mounds are exemplary air conditioners) they continue to add new tunnels, blending their work with that of other crews as they converge. Gareth Morgan describes the process as "an emerging path that works" (Morgan 1996).

Termite colonies have vast networks of communication that enable collaborations or work teams to share what they have done with others, so that a marvelous castle emerges without a blueprint or hierarchical control. With no particular outcome attached to the termites industry, patterns emerge, creating a surprising structure. Similarly, regions with very broad and diverse communication networks allowing people to form small groups to work on innovative projects can often quickly shift a region into a dynamic mode using uncertainty as a catalyst for unique and unexpected opportunities.

16.3 Transformative Entrepreneurs: Entrepreneurs that Make a Difference

In Appalachian Ohio, a regional entrepreneurship organization, ACEnet, has been formed to facilitate entrepreneurial behavior in the Appalachian region. This organization has been involved in a wide variety of efforts to bring new businesses to the region and to exploit existing business opportunities. Observations of work at ACEnet have led to many insights about the role of transformative entrepreneurs.

An entrepreneur, according to Jay Kayne, is "a person who sees an opportunity and acts to create an enterprise around that opportunity" (Kayne 2000). Having many entrepreneurs in a region is important - they can create wealth, help respond to changes in the economy by creating new businesses to replace the old, and create new jobs for the region.[3]

However, just having many entrepreneurs located in a region is not enough. Two regions may have the same number of entrepreneurs and yet one region is vital and dynamic and the other has a humdrum economy. The difference is the presence of *entrepreneurs who make a difference*. These entrepreneurs self-organize, joining with others for mutual benefit, and in the process, also benefit the area economy. The following sections describe these entrepreneurs, and then show how together they become the foundation of a healthy economic ecosystem.

[3] "A vigorous local entrepreneurial sector is almost uniformly a feature of successful [rural] areas when compared to the less successful" (Bryden).

The four types of *transformative entrepreneurs* are:

1. Social entrepreneurs who usually start the process of transformation in a region,
2. Breakthrough entrepreneurs who create innovative high growth businesses,
3. Catalyst entrepreneurs who change the way other businesses operate,
4. Local color entrepreneurs who create businesses that make a region distinctive and a great place to live and visit.

16.3.1 Social Entrepreneurs Lead the Way

Social entrepreneurs jumpstart the dynamics of change in a region. Social entrepreneurs help create *a* field of potential for entrepreneurs: a set of processes, networks, and new institutions and programs that enable many entrepreneurs to startup and then continually respond to (or create) opportunities. For a region to undergo significant change, many social entrepreneurs must emerge, and the set of social entrepreneurs will be more effective if they include a wide range of people: business entrepreneurs, agency and governmental staff, and staff or members of local economic development and community organizations. However, staff of non-profits, whose fulltime job is transforming the economy, and who have the flexibility to respond to opportunities quickly, must do most of the work.

In transforming regions, social entrepreneurs join together in small projects or collaborations, setting up key new entrepreneurial support institutions or programs that are linked in a synergistic fashion so that they begin to create a new microclimate in which much innovation and experimentation can occur. As a result, new businesses sprout. Some go on to successfully expand and create new niches. Like termites building castles, social entrepreneurs experiment and if the results work, then they continue with more projects that create new capacities.

The social entrepreneur introduces those around them to the notion of *complex reciprocity*, key to expanding resources for the region. The social entrepreneur opens resources freely, with no expectation of a direct return, but with a clear expectation that that entrepreneur will share some of their resources (knowledge, connections, etc.) with others in the network. The folkloric concept of "the gift that keeps on giving" (Hyde 1983) which continues to expand as it moves, is a very real aspect of transformation dynamics. As entrepreneurs grasp this concept, they take on mentoring relationships, share information about new markets, and volunteer to lead Q & A sessions for other entrepreneurs.

16.3.2 Breakthrough Entrepreneurs

Breakthrough entrepreneurs are making breakthroughs in products, processes, or markets, and as a result, tend to have growth rates that are much higher than the average. Breakthrough businesses create businesses that do well, not only for themselves but for the region as well (Callan and Guinet 2000). Breakthrough businesses develop new processes and/or products each year, and perhaps because of this need for continual innovation, they create jobs that pay well, are engaging and creative, and last even through downturns in the economy.[4]

16.3.3 Catalyst Entrepreneurs

Regional transformation occurs, however, only when catalyst entrepreneurs link breakthrough entrepreneurs to businesses characterized as "on the fence," that is, businesses that have a few of the characteristics described above, but are not fully committed to innovation, growth and high performance. A European Union sponsored research by Jussi Autere documented that "...if a new firm is exposed to growth oriented mental models early in its life, this is likely to install a growth-oriented ethos in the firm and lead to faster growth later in its life" (Autere 2000). If catalyst entrepreneurs can support the leadership and influence of breakthrough entrepreneurs, the region can reach the tipping point (Gladwell 2002) where the entire region becomes endlessly innovative.

16.3.4 Local Color Entrepreneurs

Regions need to identify seeds of their regional character (Austin, Texas, has its singer/ songwriter culture, Athens, Ohio, its great food), and through customized technical assistance and loan programs support new and existing businesses that reinforce and deepen that character. Other key ingredients of regional character are film festivals, antique malls, innovative trails, farmers' or other outdoor markets, and cafes. Such businesses create a *culture of innovation* that can impact the level of innovation in all sectors of the economy.

The notion of "innovative milieux" articulated by a number of European scholars points out how a region can become, in effect, an incubator for key processes such as innovation (Konstadakopulos and Christopoulos 1997). Even clusters considered low tech, such as food processing and the artisan trades, have processes of innovation "more advanced and complex than often depicted." Knowledge based, their local institutions have developed "ways of innovating, influencing the way in which knowledge is created, diffused and applied" (Bergman et al. 2001).

16.4 Supporting Transformative Businesses

The first step in supporting transformative entrepreneurs is identifying them. ACEnet has developed a set of criteria to assist staff (and entrepreneurs) to identify whether a business is a breakthrough business or a catalyst business. Basi-

[4] Meyer-Krahmer (1985) found that those involved in innovation on a regular basis had greater increases in employment during times of economic growth and smaller decreases during recessions.

cally, a breakthrough business - no matter what its age or size - has rapidly increasing sales growth; is increasing the percentage of sales in regional, national or international markets; and, has developed three or more new products or processes each year (Mairesse and Mohnene 2001). Next, ACEnet has modified its client tracking software (The Exceptional Assistant)[5] so that it can provide staff with instant reports on the percentage of time that they are spending on transformative businesses. This way, staff can self-manage their own increase of time spent with these businesses.

16.4.1 Heterarchies and Design Principles

The experience in dynamic regions has shown that once about 50 social and business entrepreneurs are networked and are self-organizing into collaborative projects, the regional economy shifts and a new state emerges, one David Stark calls a heterarchy. David Stark, in his work at the Santa Fe Institute, defines heterarchy as sets of organizations that operate with minimal hierarchy and, at the same time, have organized heterogeneity. In situations of high uncertainty, agents in a heterarchy self-organize. To prosper in such a situation, Stark believes management becomes the art of facilitating constellations that can perpetually reorganize themselves. "The solution is to minimize hierarchy," he says. "Authority is no longer delegated vertically, but emerges laterally" (Stark 1999). The goal, Stark contends, is to coordinate diverse identities without suppressing differences.

Underlying self-organized systems, whether they are immune systems, brains, termite colonies, ecosystems or healthy economies, are simple design principles.[6] Entrepreneurs and social entrepreneurs who use these principles to guide their behavior will find their businesses *and* region flourishing.

The principles are:

- Create a unique identity, but continually reinvent it,
- Consciously weave vast networks that bring in new ideas and perspectives,
- Collaborate with others for mutual benefit,
- Be surprised and use that uncertainty and surprise to make leaps in understanding.

The interactions that result from following these principles can rapidly move an economy into a highly competitive, healthy place. The new economy has qualities that are more than the sum of the parts. These design principles set off trajectories of transformation. There are dynamic processes that can dramatically accelerate economic revitalization when implemented in tandem. These are:

- Network building,
- Innovating,

[5] http://www.nextgen.on.ca/tix/en/index.cfm.

[6] In complexity science, John Holland of University of Michigan, talks about genetic algorithms - simple rules or design principles - that guide interactions among agents and enable much more complex systems to emerge.

- Clustering and collaborating,
- Making breakthroughs.

An important study done in Italy shows the relationship among these processes (Cainelli et al. 2000). The study found that even in traditional sectors such as food and textiles, innovative firms have higher labor productivity and at the same time pay their workers high wages. The innovative firms also have higher gross operating profit, Return on Investment (ROI), Return on Equity (ROE), and Return on Sales (ROS). The study also reveals the advantage of firms that network in an informal, geographically dense cluster. Both non-innovating and (even more so) innovating firms in an industrial district, when dense networks and collaborations among firms are commonplace, show the same positive performance indicators (higher productivity, labor, ROI, ROE, ROS) when compare to isolated businesses. Finally, the study also shows how over a very short period of time - 6 years - the percentage of innovators in traditional industries increased from 25 to 44 percent of all the firms in that cluster, an indication of the accelerating impact of the synergies taking place among networking, collaborating, clustering, and innovating processes.

16.5 Network Building

Networks are, simply, people with similar interests sharing information, building relationships, and working on joint projects. To the extent that networks include diverse perspectives, they can promote a continual stream of learning and innovation that help one use surprise and uncertainty as opportunities. The most basic and essential transformative process is the building of strategic networks. Each business and social entrepreneur needs to weave a world class network that contains:

- People with whom they work well,
- People who can provide new vistas, perspectives and technologies,
- People who provide specialized expertise,
- People who help open new markets,
- People who provide access to appropriate capital, and
- People who initiate collaborations (Bryant 1998).

Enhancement of innovation practices in transformed regional economies includes two processes:

- Catalyzing increased innovation activity,
- Increasing the diffusion of innovation.

Transformed regional economies have many activities and programs that increase the amount and quality of innovation that area firms and support organizations come in contact with - both from "outside" the region and "within." Social entrepreneurs need to very consciously assist entrepreneurs and their supporters to build extensive innovation milieu. The divide that currently exists between researchers in universities or labs and entrepreneurs is in most cases enormous, and bridging this gap can unleash powerful innovation forces. ACEnet linked very small specialty food processors to food scientists, product developers, and graphic designers - which led to a proliferation of high-value products.

16.6 Clustering and Collaborating

Transformation happens not through a centralized planning process but through thousands of collaborations. These range from small joint purchasing activities among sets of firms to collaborations formed to commercialize new technologies and collaborations to develop new regional infrastructure. Much regional collaboration emerges from specific clusters - sets of businesses that directly or indirectly serve the same final market.[7] Social entrepreneurs select clusters on which to focus by the entrepreneurial energy and leadership already existing in that cluster (Raines 2002).

Clusters must develop critical mass before they can manifest accelerating growth and development, so initial efforts by social entrepreneurs must include efforts to identify potential entrepreneurs in that cluster and provide support needed for them to start a successful innovative business. Once a set of cluster entrepreneurs are networked and collaborating, the cluster takes off: many more people decide to start businesses in the cluster, an explosion of collaborations appears, and some of the businesses begin to break into regional, national and international markets.

Collaborations can include policy initiatives for local, state, and federal government, joint business projects, and projects to develop entrepreneurship infrastructure. Such projects should be chosen for development based on constraints that firms identify to pursuing high-value opportunities. Projects that enhance product innovation are critical in creating clusters that function in high-value regions of the economy. Often small networks form that meet regularly to identify opportunities and needs in areas such as access to capital and markets. Often the collaboratives develop new infrastructure that is cluster specific: a product innovation award program for food businesses or a marketing initiative for tourism businesses.

16.7 Breaking Through

The new economy described here is one where people experiment to move toward a healthier economy, and as they notice and share with each other the outcomes of

[7] To clarify terms: an industrial district, the term used first to describe the vibrant sets of firms in a geographic areas, implies that these firms are already well-networked and frequently collaborating; the terms cluster, popularized by Michael Porter, is used to help economic development staff identify a set of firms among whom they then build networks and coordinate collaborations.

their actions, begin to identify *patterns of success*. This requires a multitude of breakthroughs, small and large. The social entrepreneur can use several interesting tools and strategies to encourage breakthroughs. Four major breakthrough strategies are:

- The use of diversity to produce creative abrasion,
- Setting up explicit and informal processes for learning and reflection,
- The use of Network Mapping and Analysis to enable social and business entrepreneurs to self-manage the regional network for maximum breakthroughs,
- The creation of policy networks to link patterns of success to policy formation processes.

One of the most difficult habits that social entrepreneurs need to break is an addiction to perspectives similar to one's own. It is important to learn to listen to others with different points of view and to move out of a historical viewpoint into a completely new vantage point. Looking at the situation from this new vantage point often produces explosions of insights and innovations, which are likely to lead to increased business and regional competitiveness. Transformative regions remove the wall of difference so innovations erupt and flow throughout the region.

16.8 Learning and Reflection

Both social and business entrepreneurs tend to be action oriented. However, a small commitment to set aside time to reflect on projects - while they are underway - is one of the best investments a region can make. Over and over, tremendous benefits accrue when a group of people comes together to make sense of what they have been doing. The insights obtained from such sessions can save money and endless hours by helping groups collectively agree that something is not working and needs to be stopped, by pointing out new ways to do things, and by accumulating suggestions for improvements.

16.9 Using Network Mapping and Analysis to Maximize Breakthroughs

The last few years have seen an upswing in the number of books that combine network analysis with complexity theories (Watts 1999). Many people are beginning to identify basic guidelines for optimal network formation to support transformative economies. These are network hubs and boundary spanners.

Network hubs - people with many connections to others - who are also innovators (breakthrough or catalyst businesses) enhance the rapid spread of ideas; however, having only a few network hubs increases the instability of the network, since one hub's departure would dramatically decrease the flow of information. In addition, network hubs need to buy into the culture of openness and sharing or they could become destabilizing power brokers.[8] People tend to cluster with those like themselves and, therefore, working with boundary spanners to link isolated sub-clusters can greatly increase the flow of innovation in a network.

16.10 Policy Networks

One of the biggest obstacles - and a place of tremendous opportunity - is the current state of economic development policy in general, and policy relating to entrepreneurship specifically. Not only is the content of policy a problem, but the way policy formation takes place does not fit the needs of rapidly changing economies that are characterized by uncertainty and surprise. As one European scholar pointed out:

> Policy for innovation, as much as innovation itself, is a learning process. This implies that those responsible for executing cluster policies...need to possess certain capabilities. Adopting a cluster approach implies that the policy-making process, the day-to-day operations of innovation policy makers, is changing. Policy makers will to a much lesser extent implement policy measures and policy programs top-down. They are simply part of the decision making taking place at network level. Creating favorable conditions for straightforward interaction between the parties concerned is one of the most logical outcomes. (Bergman et al. 2001)

Some guidelines for policy that supports regional transformation are:

- Rural policy needs to support interactive processes that enable people in regions to experiment, create, and then continually improve what they have created. "Cluster policy is intended as a catalyst, creating change in collective behavior that extends beyond the immediate set of policy beneficiaries" (Raines 2002).
- Policy networks need to link regions to other regions around the world so that all have access to a thick portfolio of provocative ideas and nourishing resources.
- Policy networks need to support learning processes. People who have accomplished something together must be encouraged to share their successes (and even more important, their failures) with others so that learning can occur rapidly and spread throughout the system. Policy needs to incorporate accountability through learning and improvement processes (Diez and Esteban 2000).
- Policy needs to invest in regions, rather than establish rigidly defined and implemented government services. Region s need to be supported as they develop

[8] "In those places where networks appear to be more open and inclusive and less associated with a notion of an 'elite group' or clique, the relationship with entrepreneurship appears to be positive...a strong and exclusive network structure can inhibit or prevent new entrepreneurs, and be socially divisive" (Bryden).

systems and experiments that grow from their history and meet their needs - as they identify them.

16.11 Next Steps

To learn more about regional economies, contact the Complexity, Entrepreneurship and Regional Economies Learning Cluster which is facilitated by the Plexus Institute (http://www.plexusinsitute.org) and June Holley of the Appalachian Center for Economic Networks (http://www.acenetworks.org). Please send email to juneh@acenetworks.org or call 740-592-3854.

This paper is the result of the Complexity, Entrepreneurship and Regional Economies Learning Cluster. Many thanks to Valdis Krebs, Tom Petzinger, David Bayless, Erik Pages, Chris Gibbons, Patrick Von Bargen, Henri Lipmanowicz, Lisa Kimball, and Curt Lindberg.

References

ACEnet (Appalachian Center for Economic Networks). ACEnetworks.org, OH

Autere J, Erkko A (2000) Is entrepreneurship learned? Influence of mental models on growth motivation, strategy, and growth. http://www.tuta.hut.fi/units/Isib/publications/working_papers/ja.ea.learntentrepreneurship.2000.pdf

Axelrod R, Cohen MD (1999) Harnessing complexity: organizational implications of a scientific frontier. Free Press, New York

Brown JS http://www.creatingthe21stcentury.org/jsb17-creative-abrasion-leadership.html

Bryant K (1998) Evolutionary innovation systems: their origin and emergence as a new economic paradigm. In: A new economic paradigm innovation-based evolutionary systems, discussions in science and innovation, Commonwealth of Australia

Bryden J, Hart K (2001) Dynamics of rural areas (DORA), The international comparison. http://www.abdn.ac.uk/arkleton/doradocs/icfinal.pdf

Burt R (2000) The network structure of social capital. In: Staw B, Sutton R (eds) Research in organizational behavior (22). JAI Press, New York

Burt RS (2003) Social origin of good ideas. Preprint. http://web.mit.edu/sorensen/www/SOGI.pdf

Cainelli G et al. (2000) Technological innovation and firm performance in italian traditional manufacturing sectors, innovation and enterprise creation: statistics and indicators. European Commission Publication

Callan B, Guinet J (2000) Enhancing the competitiveness of SMEs through innovation. Proceedings of the Conference for Ministers responsible for SMEs and Industry Ministers, Bologna, Italy

Diez MA, Esteban MS (2000) The evaluation of regional innovation and cluster policies: looking for new approaches. University of the Basque Country, online article

Flora JL (2001) Social capital and communities of place: social capital library. Working papers on social capital. The World Bank Group. http://www.worldbank.org/poverty/scapital/library/flora1.pdf
Flora JL et al. (1997) Entrepreneurial social infrastructure and locally-initiated economic development. Sociological Quarterly 38(4):623-645
Florida R (2002) The rise of the creative class: and how it's transforming work, leisure, community and everyday life. Basic Books, New York
Fountain JE (1997) Social capital: a key enabler of innovation in science and technology. In: Branscomb LM, Keller J (eds) Investing in innovation: toward a consensus strategy for federal technology policy. MIT Press, Cambridge
Gladwell M (2002) The tipping point. Little, Brown, and Co, New York
Hertog PD (2001) In pursuit of innovative Cclusters. Taipei, R.O.C.
Holley J, Krebs V (2001) Tracking that makes a difference. ACEnet Institute, Athens, OH
Holley J, Krebs V (2002) Opportunities: sustainable communities through network building. ACEnet Institute, Athens, OH
Hyde D (1983) The Gift: Imagination and the Erotic Life of Property. Vintage Books
Kelly K (1994) Out of control: the new biology of machines, social systems and the economic worlds. Addison-Wesley, Reading
Konstadakopulos D, Christopoulos D Innovative milieux and networks, and technological change and learning in european tegions: technology policy and innovation strategies. University of West of England, Bristol
Krebs V (1996) Visualizing human networks, Release 1.0 (February):1-25
Lin N. (2001) Social capital. Cambridge University Press, New York
Mairesse J, Mohnene P (2001) To be or not to be innovative: an exercise in measurement. Merit
Maskell P (2001) Growth and the territorial configuration of economic activity. http://www.druid.dk/conferences/nw/paper1/maskell.pdf
Meyer-Krahmer F (1997) Globalisation of R&D and technology markets: consequences for national innovation policies. Proceedings of the international conference on 1 and 2 December 1997 in Bonn, Petersburg, Germany. Technology, Innovation and Policy V(9)
Morgan G (1996) Images of organization. Sage, California
Porter M (1998) The competitive advantage of nations. Free Press, NewYork
Sabel PC (1990) The second industrial divide. Basic Books, US
Raines P (2002) The challenge of evaluating cluster behavior in economic development Policy. European Policies Research Center, UK
Rogers E (1995) The diffusion of innovations. Free Press, New York
Stark D (1999) Heterarchy. http://www.santafe.edu/sfi/publications/bulletins/buletinfall9.../organizationdiversity.htm
Stark D Collaborative organizations and interactive technologies. http://www.sociology.columbia.edu/downloads/other/dcs36/research_statement.pdf
Valente TW (1995) Network models of diffusion of innovations. Hampton Press, Cresskill
Wasserman S, Faust K (1994) Social network analysis: methods and applications. Cambridge University Press, New York
Watts D (1999) Small worlds. Princeton University Press, Princeton
Watts D (1999) Networks, dynamics, and the small-world phenomenon. American Journal of Sociology 13(2):493-527

Watts D, Strogatz S (1998) Collective dynamics of small world networks. Nature 393(6648):440-442

17 Uncertainty as Certainty

Tom Petzinger[1]

I am trying to make money in the biotech industry from complexity science. And I am doing it with inspiration that I picked up on the edge of Appalachia spending time with June Holley and ACEnet when I was a *Wall Street Journal* reporter. I took some of those ideas to Pittsburgh, in biotechnology, in a completely private setting with an economic development focus, but also with a mission to return profit to private capital. And we are doing that. I submit as a hypothesis, something we are figuring out in the post- industrial era, that business evolves. It is not the definition of business, but business critically involves the design of systems in which uncertainty is treated as a certainty. That is what I have seen and what I have tried to put into practice.

My first exposure to business was when I was eight or nine-years old. I would go to my father's business on Saturday mornings after I was done at the YMCA and I would spend an hour or two with him before he shut down at noon. My father's business was one of the early travel agencies. Something was always going wrong. A travel agent would rush into my father's office with a horrified look, burst into tears, and say "Mrs. So and So's flight was delayed; I made the wrong reservation; we never got the receipt; we didn't send the money in on time." My father was completely unflappable and said, "That happens. Now let's figure out how to solve this problem." And invariably when the tears had dried and the problems had been solved he would circle back and gently say, "What did we learn from that? What can we do to make sure next time that happens it is not such a surprise?" A simple thing. But now when I have people running into my office saying, "The university wants to take back the license"; "We didn't get the funding"; "The experiment didn't work." My response is, "What can we learn from this?" The unexpected is to be expected. This is what business is. It is the unexpected. How can we get something back from the unexpected?

In terms of how one uses this as a design criterion in business, an image that has stayed with me is visiting the 911 Center at the Phoenix Fire Department during my days as a journalist. I was in Phoenix writing a column for *The Wall Street Journal* because I had heard the Fire Chief in Phoenix had written a book called *Elements of Fire Department Customer Service*. What an odd title for a book. This Fire Chief understood complexity and uncertainty in an economic setting full of unpredictability like nobody I have ever seen. I asked him, "How do you try to design for a system that is built on never knowing where the next alarm is coming from?" He said, "Follow me," and opened the door to their 911 Center, a small, darkened room with a few dozen people sitting at consoles. There was a little bit

[1] Tom Petzinger, Chairman and CEO, LaunchCyte LLC., A Bio-Informatics Network, tom@launchcyte.com.

of murmur going on; it looked very productive and efficient, very business-like. We walked behind the people, listening to the conversations. "Your child is at the bottom of the pool?"; "Your husband is down?"; "Where is the gunshot wound?"; "We will get someone there immediately." There was crisis upon crisis upon crisis. But when you stepped back and looked, it was completely orderly, structured, and efficient.

Journalism is another system in which there is an abundance of certainty intended to contain continuous uncertainty. Journalism is a system of very rigorously followed rules. What is the deadline? Who do I get this copy to? It has to be deadly accurate; there cannot be one factual error in an article that was assigned to me 90 minutes ago. It has to be done in 90 minutes and I don't know anything about this topic. I have to know the limits of my own knowledge and how widely or narrowly, or how specifically or generally to write in order to assure that it communicates some meaningful information but is not inaccurate because I didn't know which questions to ask, much less how to spell everybody's name. That is the fundamental uncertainty, the entire grist of journalism. But it exists within a system that recognizes and embraces that everything will be uncertain. There are rules and support designed to cope with that uncertainty.

I wrote about big business for many years; business that is a machine; business in which the role of management is to create an ideal, an optimum and maintain it; business in which decision-making operated on the basis of spreadsheets or risk analysis. But after visiting the Phoenix 911 Center and Ace Net's Community Kitchen Incubator, and after thinking about how my father ran his business, I decided to get out of the press box and go into the playing field. I became a practitioner of the things that I had observed, trying to put into practice the theories that I had developed, including largely the theory of self-organization and complexity. I created an environment in the biotech industry (much like June Holley did in Appalachia) in which innovators could come and add value to their ideas, in part by collaborating with each other. Instead of putting it in a very agriculturally rich area, we did it in a very rich intellectual environment surrounded by the University of Pittsburgh, Carnegie Mellon University, the Hillman Cancer Institute, and seventeen world-class hospitals.

We decided to start licensing in technologies to create multiple companies simultaneously; just as Ace Net helped foster the creation of multiple food companies. It is a portfolio approach. Some will fail. But as they fail we will learn from their failure in a way that enriches the odds of all the others. The business is based on ideas and knowledge, not on assets. One can never predict when an idea will come. All you can do is foster the conditions in which they are likely to come. When we make investment decisions we do discount cash flow analysis and risk analysis. We look at it in the context of our portfolio. We spreadsheet ourselves to death and yet when it's all done we make gut decisions based above all on the quality of the people that we are dealing with. This is a great potential business. But what is really great about it is that the inventor or the group of faculty that is associated with it is made up of people that are adaptable, flexible, creative, and they will keep inventing. Because if you license an invention into a new company, which is what our model is, and if in this environment you don't have an idea what kinds of people are going to be able to survive in this environment, then you are doomed. On the other hand, when you invest in the ideas and knowledge of each other, then that's got real power.

I am going to close by addressing a question. "How would one expect the reaction or adaptation to a downturn to differ in a healthy organization versus an unhealthy or dysfunctional one?" These are tough times. Funding is challenging; business conditions have slowed even in biotech which has lagged the downturn in the markets. One of the things we have realized is that if you are involved in a system in which everyone has a stake in the outcome (it may not be the same stake; but it is proportional to the extent that we all win if any of us wins, or we all lose if any of us loses) then the group makes its own decisions about how to cope with challenging times. If there are pay cuts to be had; if we have to defer a deal; if on the other hand, we want to accelerate a deal because it helps our story and it is going to increase our funding likelihood, even if it increases our risk, everyone owns a piece of the good and the bad. If everyone is a stakeholder in my organization then we are all rolling in the same direction. When resources are limited, companies can deal with one another in ways that harvest or husband each others cash. They can deal with each other in kind. They can do experiments in our industry for one another, in which cash may not flow, but in which the fruits of those experiments are shared and enrich the value of each of our respective activities.

Section IV Conversations on Uncertainty and Surprise

18 Uncertainty and Surprise: Ideas from the Open Discussion

Michelle E. Jordan[1]

Approximately one hundred participants met for three days at a conference entitled "Uncertainty and Surprise: Questions on Working with the Unexpected and Unknowable." There were a diversity of conference participants ranging from researchers in the natural sciences and researchers in the social sciences (business professors, physicists, ethnographers, nursing school deans) to practitioners and executives in public policy and management (business owners, health care managers, high tech executives), all of whom had varying levels of experience and expertise in dealing with uncertainty and surprise. One group held the traditional, statistical view that uncertainty comes from variance and events that are described by usually unimodal probability law. A second group was comfortable on the one hand with phase diagrams and the phase transitions that come from systems with multi-modal distributions, and on the other hand, with deterministic chaos. A third group was comfortable with the emergent events from evolutionary processes that may not have any probability laws at all.

The diversity of conference participants was vital to the aims of the conference. Bringing together people with very different views strengthened the probability of extraordinary exploratory behavior and the hope of producing entirely new structures, capabilities, and ideas. Out of our interconnections might emerge the kind of representation of the world that none of the participants, individually, possess or could possess. Participants seemed to share a belief that in our lives and businesses we need to figure out ways to begin to communicate among various domains, forging new relationships and connections that will help us decide where we can make progress, definition, and increased understanding.

One purpose of the conference was to develop the capacity to respond to our changing science and to new ideas about the nature of the world as they relate to the unexpected and the unknowable. We hoped to identify common themes that emerge between social systems and natural systems when we consider fundamental uncertainty and to address the impacts of these new perspectives on the research agenda of the natural and social sciences. We hoped to explore the relevance of these new perspectives for practitioners and executives in public policy and management. We tried to begin to open up some new research and practice questions, identify some themes and research issues, and perhaps inspire some to pursue those research issues. Importance was not placed on how much individuals learned over the three days, but on identifying what we need to learn as the result of being there.

---

[1] Editor, McCombs School of Business, The University of Texas at Austin, majordan@austin.isd.tenet.edu

Participants were cautioned that the danger in trying to come up with a research agenda is that you shut down possibilities when you begin to define and put boundaries on problems and decide what kinds of methodologies are useful. But at the same time, just kind of wandering around doesn't help either. Participants tried to do some agenda building, while at the same time fully recognizing that part of the secret of being good at this work is to back away from that agenda and be open to how that agenda, itself, grows and develops over time.

Conference participants listened to two keynote speakers and three groups of four panelists. At the end of each panel the entire group engaged in a dialogue inspired by the presentations. Near the end of the conference participants had the opportunity to meet twice in small breakout groups of their choosing for free-flowing conversation. Each breakout group reported back to the larger group. Holding doubt, stating assumptions, and valuing uncertainty were things participants highly respected throughout the dialogue process. Participants shared a common belief that uncertainty draws us on as we learn more. We learn not so that we can answer questions or resolve uncertainty, but so that we can discover more questions.

We have made every effort to assure that all the ideas in this Chapter were ideas expressed by conference participants at the meeting; they are not the ideas of the editors or the compiler. All whole-group conversations between conference participants were recorded and transcribed, and the transcriptions were reviewed by a third party. In trying to capture the informal dialogue, we made no attempt to present the material chronologically. This chapter is not a record of what happened; rather we attempt to identify themes, ideas, tensions, issues, and insights that emerged in the conversations between conference participants. After attempting a first draft of this summary we went back, re-read the transcriptions, and made revisions. The editors of the proceedings then read the material and made suggestions. Then, we edited the summary for readability and flow of ideas. As we attempted this synthesis of the conference conversations, we noted areas of wide agreement and consensus, areas where slightly opposing viewpoints or shades of meaning caused slight disagreement, and areas of deep-felt conflict among participants. We also noted that some points were raised that seemed to be important, but were not pursued. We have tried to capture all of these interactions within this summary.

One of the greatest difficulties we had was in trying to capture the social construction process. It is fairly straightforward to capture the result of the process, but difficult to follow the process through its evolution. Therefore, we sometimes fail to capture the anxiety, hesitancy, excitement, or urgency associated with some of the ideas and conflicts. Another problem we encountered was due to the effort made during the conference to keep the conversation tentative. There were no debates, no votes, and no attempt to come to consensus. This was a deliberate decision based on the multiple aims of the conference, but that process makes it difficult to write about the results without overstating the conclusions. We end up writing from a stronger position (with less doubt and hesitancy) than the participants' exchanges actually contained.

In the remainder of this work we start with a discussion of the need for clarity in our terms and definitions as we wrestle with complexity science. Then we discuss emergence, one of the most prevalent themes throughout the conversations, which leads to a discussion of the tension between stability and fluctuation. This is followed by a discussion of the difficulties of managing individual and collective elements and specializations as a strategy for facing uncertainty. We then discuss the nature and limits of information and reason, move to responses to uncertainty and defining success in the face of uncertainty. Lastly, we explore issues of methodology.

The work contained in this summary came out of the collective mind of those gathered together. No attempt is made here to attribute contributions to individual participants. This reflects our belief, shared by many participants, that ideas are socially constructed. As was noted by one participant, "It is futile to think that you can attribute to an individual something specific when our ideas are socially formed. Certainly we have individual ability to surprise in the way we interact with each other, but to attribute some new idea to one individual disregards the socialness of what we are as individuals and as individuals together." This does not imply that all participants agree with all statements contained in these proceedings. Rather, we hope that all participants will feel some sense of ownership in these discussions.

18.1 Difficulties Talking to Each other

Participants recognized early on their difficulties in communicating with one another across the diversity of their backgrounds. One of the issues the group tried to resolve as it went forward was differences in levels of understanding and experience related to the theme of uncertainty and surprise. People from different disciplines had very deep knowledge in some areas and not particularly deep knowledge in other areas. There was a tremendous language problem; many conference participants had great difficulties in bridging the different idioms. Some of the greatest challenges facing our efforts for future integration and collaboration are the following: understanding what the relationships are between the fundamental things happening in the natural sciences and the fundamental things happening in social science; finding ways to link that through complexity, and communicating in a meaningful way.

The desire for a common language was a reoccurring theme among conference participants. How can we communicate better across disciplines? How can we create organizations that have language and structures that support a diversity of opinions? Participants discussed various structures to talk to each other. One idea was to develop a tutorial that would help everyone understand the concept of uncertainty, or a dictionary to "fine tune the language." Another idea was to create metaphors by developing side-by-side stories from natural science, social science, and practice in order to bridge between deep knowledge in different areas and identify and illuminate the connections between science and practice. Some participants called for more dialogue through creation of opportunities for conversation between scientists and practitioners or by exploring "open space" as a mechanism for creating the conditions for self-organization. Communicating through visual language (i.e., graphics, shapes, and images) was put forth as "the only way we can communicate as a body" where not all of us have the same level of mathematical sophistication.

Despite the difficulties participants saw value in trying to talk across a diversity of disciplines. The group was encouraged to steer the course and to recognize that natural science versus social science is a false dichotomy. Though we may be speaking sometimes in different languages; we are all really struggling with the same issues and thinking about the same things. Complexity science and Chaos are confronting the boundaries of science itself, going beyond traditional or stereotypical scientific ways of looking at the world. Everyone has important contributions to make to that.

18.2 What is Complexity?

Conference participants wrestled with differences in language as they tried to work out questions and ambiguities regarding even the fundamental themes of the conference, including the definitions of complexity, emergence, and uncertainty. Can we name or label what complexity is? What is the difference between calling it a theoretical framework, a worldview, and a science? With engineers, mathematicians, social scientists, and general practitioners in the room, responses ranged from "Who cares what you call it?" to "I care tremendously what you call it because that is what I do." None of the participants offered any definitive definitions of complexity, and the question remained open. Participants also questioned the uses of complexity theory. Is there utility to complexity or is it simply useful to describe a particular problem and help us make sense of it? Can we identify places where real improvements have been achieved by using complexity theory? Does it help you with pattern recognition?

The nature of uncertainty was explored as participants attempted to define the difference between fundamental uncertainty and other kinds of uncertainty. One participant offered a mathematical perspective. The simplest distinction that can be made is that we have signal plus noise, and noise is not part of the system; it is something external to the system. The more we learn about the system and the system dynamics, the more noise we can eliminate. But after we eliminate all the noise and just have that system, there is an intrinsic dynamic that has an implicit uncertainty because it is part of the dynamics of the system. That is fundamental uncertainty; and that you cannot get rid of. You can not treat it like noise because it has information in it. If you learn how to do the analysis properly, you can extract that information to your benefit.

Other participants agreed that fundamental uncertainty is the kind of uncertainty that is not resolved by more information. Studies in Chaos teach that even deterministic systems have irreducible randomness that cannot be reduced by knowing more about the system; at best we can predict where the randomness may occur. In quantum studies, at the tiniest scales, all we have is probability. In cosmology the more we learn about the universe the more questions we have about it. From a statistical perspective more information will not help you get rid of the fundamental uncertainty intrinsic in wicked decision problems There may be multiple, equally good explanations so you must interpret reality in very different ways. There is no absolute answer.

In attempting to resolve what may be fundamental uncertainties we often end up introducing uncertainties through our actions, artifacts, and tinkering. And perhaps we then can control our actions, artifacts, and tinkering and reduce that kind of uncertainty. We make trouble for ourselves by introducing our unnecessary complications, thereby creating uncertainty. Introduced uncertainties complicate things whereas fundamental knowledge complexifies things. The really bad news about fundamental uncertainty is that it is irreducible through increased knowledge. And the good news about it is that it is irreducible through increased knowledge. And the wonderful thing about that is that we never need to fear that we are going to hit the wall by learning more.

18.3 Emergence

Emergence was an idea that wove itself throughout much of the informal conversation, yet *emergence* as a term created confusion among participants. There was acknowledgment of a need to state more clearly our assumptions with regard to fixed structure versus emergence. We often assume that problems have fixed structures (i.e., strange attractors) within the framework of a highly predictable dynamic. These fixed structures may hold for a time, but emergence and structural change are broader issues over longer times.

You can define "emergent" in the normal Webster Dictionary sense simply as something changes or happens, but if you use "emergence" to mean more in the complexity sense, it implies some sort of scale shift having to do with a fundamentally different structure of the organization of interactions, or a shift in the nature of the network, or of knowing, or awareness. The double pendulum does not show emergence because there is no structural shift. Although the motion of the double pendulum is chaotic and therefore not predictable, its structure is fixed; a triple pendulum will not emerge from the double pendulum the way that a democracy might emerge from a monarchy. Dimensionality as a qualitative science is an important difference between social systems and mechanical systems because social systems are more complicated in that you probably don't know the dimensions of the problem you are in; and you don't know where it's going from here.

Some participants felt that emergence is related to uncertainty in that uncertainty comes from the emergence of new things. Fundamental uncertainty is something that emerges that nobody ever dreamed of. The evolution of disease over the 20th century gives us an example of emergence through new capabilities and skills. Infectious disease was the major problem that accounted for mortality at the early part of the 20th century. As infectious diseases such as tuberculosis came under more control and became less of a problem, many common viral infections became a problem. We began to see the emergence of chronic diseases such as diabetes. Back in the 1920's people with diabetes died. With the advent of insulin they stayed alive but they began to develop problems with blindness, kidney failure, and theropathy. Those were all emergent problems. In a related issue, as we see the emergence of the capability of dealing with disease we also see the emergence of practitioners who deal more with body parts than they do with whole human beings. General practice used to be very in tune to the concerns and the human aspects of people because that is all they could do. They did not have a very powerful disease model that would allow them to do many of the things that we are presently doing. The relationship between the illness and the disease and the ability to bring them together is a complex and emerging problem for physicians.

Some conference participants felt that the conditions for emergence may include critical mass. "There is a certain place where things do take off, and they really get out of control." Developing certain kinds of business relationships might not work with just a half dozen businesses; or even a dozen. Things may begin to change after there are fifty. But simply the drive for numbers is insufficient; there is a qualitative aspect to critical mass that is about the relationships created as people get to know each other through meaningful interactions so that they can self-organize these collaborations. This requires agents who are both inside the immediate organization and outside; it is the interaction with the outsiders that helps drive toward critical mass. It is also possible to go super critical in which case, essentially, you are shutting things down.

Critical mass is also about how the picture of what is happening gets out there into the public and then moves into people's minds. Even if you have business networks interacting and networking but they don't have a notion that there is something about transformation of the regional economy or the community, nothing really interesting happens. Intelligence (wisdom) can be an emergent outcome from a process of simulation, connectivity, and consistency in a networked group. By talking together and exchanging ideas we can create that kind of emergence. There are processes you can use with organizations that tend to be reliable in terms of the quality of the outcome or the quality of what is being produced. Processes that involve a high degree of participation from people all involved together in a rich process to contribute to figuring out how to deal with something or find the answer to a particular question tend to be much more reliable, even though you may have more uncertainty than those processes that are very controlled.

Some conference participants cautioned the group not to equate emergence with miraculous magic. Despite our tendency to usually look at emergent systems as the greater coming out of the less, it is a mistake to assume that emergent properties are always positive. For example, part of our failure to deal with depression may stem from our failure to recognize it as a complex emergent system. If so, trying to find the magic bullet for depression is futile.

18.4 Tendencies toward Stability, Tendencies toward Variance

Conference participants recognized that there are tendencies towards stability and tendencies towards variance and that we may be confused about the relationship between them. Discussions highlighted tension between the belief and desire for stability and permanence (strange attractors, steady states, equilibrium) and the belief and recognition that things are not permanent and stable (fundamental uncertainty, emergence). Assumption of the need for stability could be seen in the number of questions and comments concerning systems assumed to be in equilibrium.

Our assumption about the value of stability may lead us to our assumption of the value of permanence. In the business world there is a belief that we need to design an organization that can survive and adjust to anything that comes along as opposed to being an optimizer in a certain situation and then dying off. One participant suggested that we shouldn't be so attached to our organizations; we should allow them to break down and reform when the next problem arises to form around, much like project teams do. There is evidence that the value of permanence may be a socially constructed Western trap that is not shared by Eastern philosophies. In the Philippines when people are about to die they come home and are taken care of during the last few days. They die with their families acknowledging the impermanence of life. In the United States most of the money in healthcare is spent in the last few days. The financial crisis in healthcare has a huge amount to do with a uniquely Western and particularly North American attitude towards death and impermanence.

Complexity science leads us to understand that the degree of variability in the distribution of fluctuations in system dynamics is more important than any average quantity, which is counter to the traditional paradigms of medicine, management, and scientific research. We used to believe that equilibrium was the optimal state for systems. Complexity science leads us to believe that stability is death and that survivability is in variability. Health versus illness was a metaphor used throughout the conference to describe phenomena at many levels, from companies to individuals, from conditions such as autism to systems such as brains and hearts. Health can be identified with scanning, searching, seeking; a constant "chaotic search." Illness is associated with being stuck or locked in, closing down, or following only one lead.

It has been known for a hundred years that each person has an average gait and a three or four percent variation around his or her stride interval. People have systematically dismissed the variation as noise because it is so small, but it turns out that all the information is in the variation. The average stride has very little information about the motor control system, but if you look at the fluctuations you can find out a great deal about the motor control system because the fluctuations are very sensitive measures about what is going on, especially in pathological situations. That theme kept reoccurring. In measuring a person's health it is not the average heart interval that is important; it is the variations in the heart interval. It is not your blood pressure that is important; it is the fluctuations of the blood pressure. Brain waves fluctuate in healthy human brains. The healthy human brain is always in sort of a chaotic search; scanning through its mind and environment because it is trying to survive. A sick brain is one that is narrowly focused with very regular brain waves. Loss of variability is an indicator towards the system's death. The way we take care of patients we don't usually measure in that way so we don't notice because we're not looking.

There seems to be an analogy between how the mind functions and how a company might function. Traditionally we shut down possibilities too early, but in order for a company to stay competitive you need humans in the company who are willing to scan. The Homeland Security Department is as an example of a system with tremendous potential for healthy interaction because it has fantastic diversities needed for scanning and sensing. But these diverse agents are not connected enough and there are not strong norms for conflict to be carried on and sustained for a period of time. Creative potential gets muted, under-exploited, or under-used in their drive towards consensus. Many organizational disasters have been situations where rookies and newcomers saw unfolding signatures of trouble. But that information never got introduced either because the newcomers were fearful or were told to shut up or the senior people didn't listen.

The tension between stability and variability is similar to the tension in the social sciences between exploitation and exploration. We often think of exploitation as a strategy for maintaining stability and exploration as a strategy for exploiting variability. We may need a balance between exploration and exploitation, stability and variability, convergence and divergence within a state. One participant explained a model in which the problem was how fishermen have to be non-rational because if they are too rational they get to the place where the fish are and they never can discover fish elsewhere. They must be exploiting fields they already know have fish while they at the same exploring new fields for fish. Fishermen have the ability to discharge the potency of the attractor by exploring, and then go on to other attractors that still exist in the mind.

18.5 Difficulties in Managing Individual and Collective Elements

An issue that resurfaced several times throughout the conference was the relationship between individual elements and collective elements. Is the individual or the organization more likely to generate a surprise? Is an organization inherently more or less knowable and under what conditions than the sum of the individuals within it? What are the implications of the individual as complex? What is the interaction of the individual and the collective? These questions emerged from the conversations of the conference participants and often caused disagreements and sometimes conflict.

Traditionally Western thought has tended towards the individual over the collective and this creates pressure towards autonomy. The opposite view is often taken by Eastern thought. Rather than thinking the collective emerges out of the individuals; Buddhist traditions teaches that each individual emerges out of the collective. "I am; because we are." Some participants felt that these lines of thought created a false dichotomy. It is not a question of *either* the individual *or* the collective, but the interaction of the two that is needed. Individual versus collective is actually a boundary issue; the individual and the group are the singular and the plural of the same process.

Our assumptions about individual and collective elements guide our ideas about collaboration, a subject upon which participants had a fair amount of disagreement. One participant stated that collaboration is the unnatural act between consenting adults that doesn't usually work very well unless there are incentives to force or encourage that collaboration. Another participant took issue with the premise that collaboration requires force to occur. There is neuro-chemical evidence that we get reward for collaborating. Brain activity occurs with collaboration that correlates with the pleasure center. If we think of collaboration as something that needs to be forced we are going to behave differently than if we think of collaboration as something to be enhanced because we want to do it. Managers need to create and regenerate a spirit where great people want to be together and contribute together, multiplying the output capability.

A conflict arose over the advisability and possibility of attribution as participants discussed incentive programs based on individual contribution to a measurable outcome or process upon which people are collaborating. Some participants felt that if we all agree on what percent of the contribution is held by each person, we can use a mathematical, financial model of incentives to share the profit, benefit, or bonus based on attributable contributions. Other participants claimed that it is futile to think that you can isolate individual contribution because our ideas are socially formed. Certainly we have individual ability to surprise in the way we interact with each other, but to attribute some new idea to one individual disregards the social ness of what we are as individuals and individuals together. Other participants disagreed, saying the individual is unique and brings something unique outside of the process, people, and interaction. Therefore, we can isolate what any individual does outside of a process with other people. There may be value in trying to identify some of the people who act as levers, amplifiers, or catalysts in particular ways; who is bringing innovation into a system and who is spreading it around. Outcomes can be changed by giving everyone real-time information about that.

In order to honor the tension between the individual elements and the collective elements, a good model might be "If you win, I win; if you lose, I lose." One participant felt that you can design an organization in such a way that people profited or lost together based upon how well they all did. One of our best levers for facing uncertainty and surprise might be to encourage quasi-autonomy (individuality) but at the same time willingness to cooperate across disciplines because this kind of collaboration gives us more capabilities and skills. By building multiple frames of reference into our organizational designs we recognize the importance of capabilities and skills present in each individual, but also the importance of the collective vision of the individuals in dealing with uncertainty and surprise. Much child abuse was not discovered for many years even though evidence was everywhere, because pediatricians and radiologists did not know what to do if they diagnosed the parents as harming their children. It wasn't until treatment teams in Boulder, Colorado, added a social worker that "Battered-child Syndrome" began to be diagnosed. It took the introduction of a new kind of skill. The capabilities that you do or do not have on the team really affect what you see and what you are able to see. And the things you see, you are able to do something about. When dealing with surprise it is important that you have your own expertise and be respectful of and learn from the expertise around you and defer to it because you are going to need help to see more of the situation.

The interaction of multiple levels or frames of reference often creates complexity leading to uncertainty and surprise, but multiple frames of reference through interacting teams of specialists can also help us deal with that complexity. The distinction between illness and disease arises as practitioners and academics become more aware of the multi-variant nature of disease and its connectiveness to multiple levels of systems. Illness is culturally developed and derived. It is how people respond to being sick and the affects the sickness has on them. Disease is how physicians recast that in terms of abnormalities of structure, abnormalities of physiology, and causal factors. Treating disease and responding to illness are both incredibly complex activities and moving from a model of disease to a model of illness is insufficient to solve the complexity issue. But for a physician to deal with the illness world in addition to the disease world seems a nearly impossible position because he or she is trying to hold two frameworks. A better response to the complexity of the relationship of these different levels may be to build teams of people with diverse specializations because the collective has the capacity to deal with greater complexity than the individual. Specialties and team treatment of illnesses involve multiple practitioners with multiple perspectives dealing with different parts of different systems in order to treat that illness.

18.6 Responses to Uncertainty

In the business world and in our personal lives surprise happens all the time. Yet even when they face the same uncertainties again and again people's response is to not expect the uncertainty, to not see that it is there, and to hope that it goes away. People seem instinctively nervous about uncertainty and this leads to pressure to create certainty. Throughout the conference participants referred to this pressure in research, businesses, and public institutions, in personal relationships and situations. This pressure to create certainty causes people to act in certain ways. People often handle the unexpected by normalizing it out of existence. They ignore questions, throw out outlying data, and assume simple solutions where none exist. People create false certainty by assuming normal distribution; but the data set about the real world does not have a normal distribution. Instead it often has a very long tail and that is where the surprise comes in because these distributions are dominated by outliers. In non-normal distributions there is a much higher probability of an unusual event occurring.

In order to weaken or neutralize the tendency to normalize and thereby change our responses to surprise and uncertainty we have to encourage things like ambivalence, questioning your associates, arguing with them even if the paradigm is underdeveloped. We must learn to think in a more mindful, less automatic manner and engage in less controlled thinking. It is more useful to talk more about embracing, accepting, and dealing with uncertainty rather than trying to reduce it; rolling with the punches, keeping it going, and coming back to the action.

In the face of uncertainty our response is often to close down on one possibility too soon. But at what point do you start to consider that you have evidence and to use it to reject possibilities? We tend to have a brainstorm, close it down to one concept, and then it has to work because that is all we have. If you build a product that has multiple attributes, they are coupled in the design. If you change several of the parameters which affect performance, you don't know which ones will have what affects. So you are in a rugged space, possibly about to jump off a cliff or find a new peak. But you really can't predict it. How broad do you search, and how deep? If you search broad and shallow, then you may find an entirely different peak that could offer fantastic possibilities. But if you search deep you are more likely to maximize your benefit from the hill you are on. You can not afford to build a prototype of every possible point in the design space, but you don't know which one design is going to work. Companies vary in their response to this complicated balance. Toyota tends to search broad, meaning that they pay for three different concepts to be explored all the way through prototyping. That lets them switch at the last minute if they need to, and they often find that they do.

Institutions may succumb to this tendency to create certainty even more than individuals do. As you move to complexity thinking you often end up wanting to move from point estimates of a variable to a probability distribution. Individual physicians seem willing to embrace this method of studying uncertainty surrounding health research results, effectiveness of therapies, or dangers of exposures. The CDC, on the other hand, does estimates, reports, and numbers to four significant figures that are likely wrong by a factor of two and does not want to hear anything about it. This may reflect some self-similarity across levels. Even with those responsible for more macro-level and more-multiple systems rather than subsystems, you still see that same pressure to create certainty being enacted at that level. Despite the fact they have a broader view, their task, their reward structure, and their incentives mirror those of people at lower levels that actually may see more clearly the complexity and the uncertainty.

Historically academic education as an institution has tended to concentrate on empirical knowledge and technical skills as a strategy for controlling uncertainty and surprise. Complexity science leads us to believe that educators must shift their emphasis to encourage students to develop the personal qualities needed to deal with surprise that include an emphasis on capabilities, experience, and skill. We need more openness in allowing students to make choices and take responsibility for trying things and failing because people have to practice how they handle responsibility and respond to novel situations. Traditionally the faculty decides what students should take and tries to make them take it. That is not good practice for learning to face an uncertain world. In giving students choices, we must be clear about our timeframe for education and that we can either offer them preparation for next year or education for a lifetime.

18.7 Defining Success in the Face of Surprise

A question emerged from the conference discussions, "What are the conditions that create success?" How do some organizations such as universities or businesses create an environment where they are creative, while others create an environment where they suppress creativity? The fact is that we don't know. People write books about companies being wondrously successful; and saying "This is it. You just have to be 3M; you just have to be Enron." But if you look through the companies that have been listed as absolutely excellent and wonderful, they later are no longer so excellent or so wonderful and may even have died.

Performance measures of an intervention are problematic because they bring the question, "What is success and what are the dimensions of success?" It is important to know that outcomes are multi-variant, meaning that there is likely to be more than one criterion for success. The crucible for understanding about multi-variant medical outcomes in the last half the 20th century has been women's health where there is a variety of what might be viewed as positive outcomes. Profitability of a product is a legitimate but limited metric for medical outcomes as are symptom base, functionality, and quality of life. Even survival itself may be a limited metric for success. One participant noted that we often make the assumption that a company must survive in order to be called successful, but success may be fleeting.

Complexity science is teaching us that success is a surprise that may not be within our control because planning and outcome are not cause and effect. A number of factors can lead to a single outcome and the interactions among those different factors can dynamically make the situation complex. From an organizational management view, decisions made in the face of uncertainty may have outcomes that are not related to planning. Organizations have to build incentive and accountability systems that recognize good decisions may have bad outcomes. Evaluation and rewards should be based on the decisions people make rather than the outcome of those decisions. Managers must look at the information that was available at the time that the decision was made and they must look at the decision process. A successful manager is not one who can control what happens; rather a successful manager is one who leverages what *did* happen in a way that is profitable to the firm.

In the organizational literature responses to uncertainty tend to be associated with issues that have answers to them and to treat the uncertainty in these issues as ignorance and error. It is assumed that if you get the system right you will be able to reduce uncertainty. A complexity definition of success cannot include freedom from error because error is inevitable. We sometimes say that if an employee in a company is surprised then he is not doing his job, reflecting our belief that information leads to knowing. But as one participant stated, "The knowing is in the doing." You cannot learn lessons from something ahead of time; you can only learn a lesson from something after you do it. Therefore we should avoid placing blame and instead learn from experience. Learning requires exploration and a "lack of constraint to wondering." Evolution had no intent except to explore. Evolution does not plan for success; it succeeds by retaining the decisions that turn out to be preferentially successful. Rather than looking at a system operating now in a stable state and trying to explain why it is so clever, we have to see evolution as a long process requiring exploration.

18.8 The Nature and Limits of Information, Reason and Rationality

Traditionally the reduction of uncertainty has been tied to information creation, but the study of complexity leads one to question the uses of information and even the definition of information. Conversations between conference participants highlighted a need to define the nature of information from a complexity standpoint. Are energy and information different constructs or the same constructs? What are the limits to reason, rationality, and knowledge? Is information a real physical entity or is it created by an observer? Is information uncertainty reduction, or does information lead to knowledge construction which may lead to understanding or knowledge that makes you less certain?

One participant felt that information must be specific to the coordinated state of the organism or how that organism is connecting to the world. Information creation is in the metastable state not in the phase dynamics, thus the phase is a key aspect of meaningful information. The brain also shows this metastable behavior. Metastability or temporary synchronization of the brain has to do with learning. A binding event makes this activity that is happening in different areas of the brain stick together for future reference so that you create some new patterns of learning that you will remember.

Conference participants debated the physicality of information. One participant drew a parallel between the construct of energy and the construct of information, saying information today is very much like the notion that energy was in the 19th century. Notions of energy had not been fully formulated; people argued and speculated about it, but it was a real, physical phenomenon. You can go back and trace information to the fundamental properties of a system and have a quantitative measure of that real, physical entity much in the same way we talk about energy today. The creation of information in quantum theoretics requires an observer. Schrödinger said the biggest mistake was separating the tree from the perception of the tree. We do not know there is a tree there until it is measured; the real physical and the observed are the same thing. H.S. Green argued that in a quantum system the information is spread out; only when you measure do you actually identify the physical entity.

In their attempts to define information, conference participants questioned the limits of information, reason, and knowledge. One of the contributions that complexity theory has made is to burst the bubble of information processing, unveiling the tentative, imperfection of knowledge. Complexity theory has shifted uncertainty toward issues of ontology rather than epistemological issues, introducing alternatives to more simple-minded reduction sort of strategies of calibration and validation. The evolutionary process starts with the evolution of emotion, only then you get the evolution of cognition and reason and sense making which can never dominate because the situation requires more than reason. So there is bound to be some kind of mix of what looks like intention, emotion, and cognition, and we flip around between these. Conversations highlighted the ambiguities surrounding questions of how different aspects of human behavior such as emotion, self-awareness, and intention influence the unfolding dynamics of human systems but cannot control the dynamics or outcome of a system or situation. Which of these aspects of human experience do we need to put in models? Can we put them in models without modeling the world? How do we fit what we know about emotion and irrationality into our understandings of complexity science?

Experiments show that intentions can stabilize and destabilize coordination states. We are born with a huge repertoire of actions. Infants can kick; they can suck; they can grasp, and so on. These are spontaneous motions so you can see immediately the possibilities of self-organizing intent. At some point, usually around four to five months, the infant knows that its actions are not produced by an external device or some machine; they actually belong to the infant. There is a transition when the baby realizes "I am directing that motion." The possibility exists that one can go from this spontaneity into agency even in more complex orders like language and society.

Self-reflexivity also has evolutionary importance because a thing that can think about itself has evolutionary advantage. Biological evolution is measured in terms of birth rates and death rates, so the only way for large, slow-reproducing organisms to compete with fast-reproducing bacteria is to replace biological evolution with cognitive evolution. Reflexivity provides totalogical arguments about the advantage of the thing that can see its own situation, and judge the dangers.

Though cognitive neuroscience has largely ignored the effect and emotional aspects of things, self-reflexivity and emotion interact to influence system dynamics and create the circumstances for surprise. Connections have been discovered between the cognitive and the limbic system where emotions and thinking begin to emerge. Syncopation recruits an entirely different neuro-network than synchronization. The brain centers that are stimulated when music is played mechanically are totally different than the brain centers that are stimulated when music is played expressively, bringing forth emotion. These are context sensitive effects. There is evidence also that stressing the system modifies dynamics and behavior. Experiments done at Harvard Medical School by Gary Holberger indicated that the time series for the fluctuations in human gait are fractal. But when they asked the group to synchronize their gait with a metronome, it changed completely due to the conscious stress you put on the system by trying to maintain that cadence. There is also evidence that uncertainty that is personally affecting, subjective rather than objective, may be particularly difficult to handle.

18.9 Research Methods and Measures

Complexity science gives us new methodologies, new ways to think about old methodologies, and new questions about evidence. Is it okay to use metaphorical and non-mathematical interpretations of complexity theory? How do you evaluate methodology? How do you weigh evidence? Are there simple rules of change that we can identify? The scientific community knew, to some extent, what evidence was after Newton's 'The Principia' was published. With new scientific thought, we are no longer so sure now what it is. We are working at the edges of things that have been known before. Conference speakers used multiple sources of evidence including data from experimental situations, simulation models, photographs of chemical reactions, and attractors (manipulation of the data in phase space). How do you compare and contrast these? How do you know if they contradict? How do you have a hierarchy of evidence?

Conference participants recognized distinctions and common denominators in the broad range of research methodologies represented among themselves and they questioned present assumptions about hierarchies of methodologies. One of the dimensions is the range of tools. Going from least precise to most precise, practitioners often use experience (anecdote), researchers sometimes use weak metaphorical connection (rugged landscape); some use stronger metaphorical connections (strange attractors); and some use mathematical modeling. Participants also recognized a range of phenomena or levels of analysis. Some participants prefer to look at surface patterns and structures. The second level is somewhat more explicit with a deeper structure available to some people. The third level of description seeks deep structures not visible without tools like time series analysis, dimension reconstruction, and high-level dimensions. These are not separate sets but continuous. From the work of the mathematicians and physical scientists who care about dimensionless invariance to the practitioners who look for ways to deal with surface structures, there is a whole range in which each of us has a place to stand and something to contribute.

Traditionally researchers believed we could look at systems in the moment to understand them, explain them, and then do things about them. Complexity science leads us to believe that a state of functionality now does not imply a state of functionality in the future. If you see trends and parameters there may be a measure of predictability about that, but cross-sectional understandings are extremely dangerous because they are the result of evolution and they are part of ongoing evolution. Therefore you should always consider longitudinal studies of things. You need to look at more than the average. The whole probability of distribution and process is governed by all the moments. The average is just one. Information is contained in higher moments like the fluctuations. Participants agreed that measures are useful; data is useful in trying to understand the problem, look for patterns, and so on. What kind of measures would help us detect signals and emerging patterns in a system? Measures that might help us look for uncertainty or instability. What measures would actually help us improve the system?

We should be carefully state our assumptions when we're talking about theories and models. The real essence of models is the dialogue with the system and to formulate and state your assumptions. When expectations concur and in some ways reinforce what is going on then you think that's okay. But when in some ways you get deviation from expectation you learn something because you say, "Obviously there is something wrong with the fundamentals of which I am assuming for this problem."

The level of description in a model is a difficult problem. Complex systems occur with different levels of description. The level of description above a given model is a kind of protective shell of the level below. The level below can wander quite a bit exploring and crossing valleys. If the phenotype still works then selection can't get through to the genotype. It can only get through when the genotype changes the phenotypes significantly. This gives you the freedom to explore and do something quite different so it can be a big jump rather than a marginal jump.

Historically in order to understand things we have looked at more and more variables. Complexity is teaching us that this may not lead to greater understanding because of the nature of complex adaptive systems such as non-linear relationships between and within variables. Participants disagreed about how much of the system you have to look at in order to understand something about the system. One participant said it would be better to start off with a simple model where you state everything and the simplification that you are making than to start off with including very contextual details so you are just describing. One participant felt that you need many sources of evidence and you need to be looking deeply into one as well. It is not an either or, but that each is built into the other. Another participant took the opposite point of view. Sometimes for complex systems you need to just look at one thing and it could reveal connections you would never think about. For example, a new method to diagnose diabetes is to look at the rumbling of people's bowels. Looking at one time series can help you understand the behavior of a whole complex system. Another participant felt that in a complex system everything is non-linearly interactive with everything else, so whatever you choose to measure is arbitrary. But the things you choose to measure should have the information about all the freedoms that you don't measure. One of the interesting aspects of complexity is that if you are sufficiently clever you might be able to determine from one long time series how many variables you really need to characterize that complex system. Another useful strategy in terms of identifying the relevant variables, because the space is huge, is to look for where there is qualitative change. To the extent you think you have a few variables that change qualitatively (because often they are the ones that are carrying the relevance) you can then map that onto dynamics. Toy models with relatively few degrees of freedom may help social scientists understand complex systems.

18.10 Narratives and Metaphors

Participants discussed storytelling as a means for understanding uncertainty and surprise. Conference participants felt that complexity science has changed and added to the place of the narrative construction of knowledge. For example ethnographers do not do statistical analysis. But they may find complexity a useful analytical framework to talk about non-linear effects of interactions among diverse agents when talking about such things as managers going from one meeting to another, presenting information to each other at different levels. But many participants agreed that researchers using this method must be cautious. A complexity story must be told in a complexity way, not in a linear way. When examining a process that was followed in order to make a decision, a complexity storyteller must look for alternative ways that it could have been approached in order to find ideas that come out of complexity thinking. A complexity narrative should have a different fundamental form than a linear, traditional narrative. It should start with a different set of concepts, a different art of story telling, in order to emerge as something appropriate. One of the problems in complexity science is that it is a form of experience; it is perceptual. The perceptual and experiential levels are vital because the experiential levels and the conceptual levels totally grow out of each other.

Conference participants entertained the question of whether it's ok to use metaphorical and non-mathematical interpretations of complexity theory and came to a cautious "yes" conclusion. One problem with metaphors is that they are often used wrongly. For example the Heisenberg Uncertainty Principle is sometimes used as a metaphor. A common phrase of "quantum jump" is often used inappropriately because quantum jumps are normally so small that you really can't see them very well. A rich area of other physical systems could be used as metaphors for social sciences and it also works in a reverse way. Physics has historically concentrated on simple, isolated systems, but as physics begins to deal a little bit more with complicated systems metaphors from the biological and social sciences become more useful in thinking about things like forest fires and self-organizing critical systems.

Evaluating narratives and metaphors as research methodology is difficult. With no mathematical modeling, you might evaluate the rigor of the construction of the metaphor in a qualitative study by looking to see if someone has used the concepts in a meaningful way. If they throw around a lot of words, don't really understand the concepts, and are using a lot of "catch phrases" without explaining what the catch phrases mean, then it is not an effective use of this methodology.

18.11 Summary

This Chapter is a report of conversations engaged in by conference participants. These included four conversations held after panels and two free-flowing breakout groups. The following themes, issues, and questions have been identified and discussed.

Language barriers and different levels of understanding created difficulties in communicating across a diversity of backgrounds.

There is confusion about the nature of complexity and fundamental uncertainty. Participants agreed that there are irreducible uncertainties that cannot be eliminated by knowing more.

Tension exists between issues relevant to systems in steady states and systems that have emergent properties. Emergence involves structural change and may require critical mass.

There are tendencies toward stability and tendencies toward variability. We often assume the value of stability but not variability. Complexity science leads us to understand that variance is more important than average quality.

There are tensions between individual and collective elements that must be managed. One of our best levers for facing uncertainty and surprise might be to encourage quasi-autonomy but at the same time willingness to cooperate across disciplines through designs such as teams of interacting specialists.

Responses to uncertainty and surprise have traditionally been to normalize it out of existence. Complexity science is teaching us that we need to acknowledge and embrace uncertainty and surprise.

Complexity science requires new definitions of success, emphasizing that outcomes are multi-variant and multifactoral. Planning and outcome may not be cause and effect.

Complexity science leads ups to question the nature and limits of information, reason, and rationality. We are also called to recognize the role of emotion, self-awareness, and intent in system dynamics.

Complexity science gives us new methodologies, new ways to think about old methodologies, and new questions about evidence. The creation of these new methodologies is in itself a serious, scholarly task.

Complexity science has changed and added to the place of narrative and metaphor as methodologies for understanding uncertainty and surprise. We must be careful to construct these non-mathematical models with care and rigor.

Understanding Complex Systems

Edited by J.A. Scott Kelso

Kerner, B.S.
The Physics of Traffic: Empirical Freeway Pattern Features, Engineering Applications, and Theory
682 p. 2005 [3-540-20716-3]

Kleidon, A.; Lorenz, R.D. (Eds.)
Non-equilibrium Thermodynamics and the Production of Entropy: Life, Earth, and Beyond
260 p. 2005 [3-540-22495-5]

Jirsa, V.A.; Kelso, J.A.Scott. (Eds.)
Coordination Dynamics: Issues and Trends
272 p. 2004 [3-540-20323-0]

MIX
Papier aus verantwortungsvollen Quellen
Paper from responsible sources
FSC® C105338

FSC
www.fsc.org

If you have any concerns about our products,
you can contact us on
ProductSafety@springernature.com

In case Publisher is established outside the EU,
the EU authorized representative is:
Springer Nature Customer Service Center GmbH
Europaplatz 3, 69115 Heidelberg, Germany

Printed by Libri Plureos GmbH
in Hamburg, Germany